COOPERATIVE EXTENSION

Forage Production for Pasture-Based Livestock Production

Edited by:
Edward B. Rayburn, Extension Forage Agronomist
West Virginia University

Written by:

A. Ozzie Abaye
Thomas J. Basden
Robert A. Byers
E. Ann Clark
James B. Cropper
William S. Curran
James T. Green
Marvin H. Hall
Heather Karsten
Badruddin Ali Khan
William M. Murphy
Lloyd B. Owens
Barbara W. Pennypacker
Paul R. Peterson
Edward B. Rayburn
Ronald R. Schnabel
Andrew N. Sharpley
William L. Stout, Jr.
Benjamin F. Tracy
Lester R. Vough

Natural Resource, Agriculture, and Engineering Service (NRAES)
Cooperative Extension • P.O. Box 4557 • Ithaca, New York 14852-4557

NRAES–172
December 2006

ISBN-13: 978-1-933395-00-5
ISBN-10: 1-933395-00-1

Library of Congress Cataloging-in-Publication Data

Forage production for pasture-based livestock production / edited by Edward B. Rayburn ; written by A. O. Abaye ... [et al.].
p. cm.— (NRAES ; 172)
Includes bibliographical references.
ISBN-13: 978-1-933395-00-5 (pbk.)
1. Forage plants. 2. Pastoral systems. I. Rayburn, Edward B. II. Abaye, A. O. (Azenegashe Ozzie) III. NRAES (Series) ; 172.
SB193.F627 2006
633.2—dc22

2006002253

Natural Resource, Agriculture, and Engineering Service (NRAES)
Cooperative Extension • P.O. Box 4557
Ithaca, New York 14852-4557
Phone: (607) 255-7654 • Fax: (607) 254-8770
E-mail: NRAES@CORNELL.EDU • Web site: WWW.NRAES.ORG

Table of Contents

ABOUT THE AUTHORS

CHAPTER 1: PLANT MORPHOLOGY AND ITS EFFECTS ON MANAGEMENT

A. Ozzie Abaye
Associate Professor
Department of Crop and Soil Environmental Sciences
Virginia Polytechnic Institute and State University

James T. Green
Professor of Crop Science
Department of Crop Science
North Carolina State University

Edward B. Rayburn
Extension Forage Agronomist
West Virginia University

CHAPTER 2: ECOLOGY OF PLANT COMMUNITIES IN FORAGE-LIVESTOCK SYSTEMS

E. Ann Clark
Associate Professor
Department of Plant Agriculture
University of Guelph

Heather Karsten
Associate Professor of Crop Production/Ecology
Department of Crop and Soil Sciences
The Pennsylvania State University

William M. Murphy
Department of Plant and Soil Science (retired)
University of Vermont

Benjamin F. Tracy
Assistant Professor of Agroecology
Department of Crop Sciences
University of Illinois at Urbana-Champaign

CHAPTER 3: SOILS, SOIL FERTILITY, AND FERTILIZERS

Edward B. Rayburn
Extension Forage Agronomist
West Virginia University

Thomas J. Basden
Extension Specialist–Nutrient Management
West Virginia University

CHAPTER 4: NUTRIENT MANAGEMENT IN FORAGE-LIVESTOCK SYSTEMS

Paul R. Peterson
Assistant Professor and Extension Agronomist
Department of Agronomy and Plant Genetics
University of Minnesota

James B. Cropper
Forage Agronomist
U.S. Department of Agriculture
Natural Resources Conservation Service
North Carolina

Edward B. Rayburn
Extension Forage Agronomist
West Virginia University

William L. Stout, Jr.†
Pasture Research Lab
U.S. Department of Agriculture
Agricultural Research Service
Pennsylvania

CHAPTER 5: ENVIRONMENTAL IMPACTS OF GRAZING

William L. Stout, Jr.†
Pasture Research Lab
U.S. Department of Agriculture
Agricultural Research Service
Pennsylvania

Andrew N. Sharpley
Soil Scientist
U.S. Department of Agriculture
Agricultural Research Service
Pennsylvania

Lloyd B. Owens
Soil Scientist
U.S. Department of Agriculture
Natural Resources Conservation Service
Ohio

James B. Cropper
Forage Agronomist
U.S. Department of Agriculture
Natural Resources Conservation Service
North Carolina

Ronald R. Schnabel†
Pasture Research Lab
U.S. Department of Agriculture
Agricultural Research Service
Pennsylvania

CHAPTER 6: INVERTEBRATE PESTS, WEEDS, AND DISEASES OF FORAGE-LIVESTOCK SYSTEMS

Robert A. Byers
Department of Entomology
The Pennsylvania State University

William S. Curran
Professor of Weed Science
Department of Crop and Soil Sciences
The Pennsylvania State University

Barbara W. Pennypacker
Adjunct Senior Scientist and
Professor of Agronomy
Department of Crop and Soil Sciences
The Pennsylvania State University

Badruddin Ali Khan
Alberta Ag, Food, and Rural Development
Beef Branch

CHAPTER 7: ESTABLISHING FORAGE STANDS

Lester R. Vough
Forage Crops Extension Specialist
University of Maryland

Marvin H. Hall
Professor of Forage Management
Department of Crop and Soil Sciences
The Pennsylvania State University

Edward B. Rayburn
Extension Forage Agronomist
West Virginia University

† Deceased

ACKNOWLEDGMENTS

The authors of chapter 3 would like to acknowledge the use of information from the following faculty at West Virginia University:

D. K. Bhumbla. *Interpreting the West Virginia Soil Test Report.* West Virginia University Extension Service Fact Sheet. HTTP://WWW.CAF.WVU.EDU/~FORAGE/SOILTEST/INTRSOIL.HTM

Willem A. vanEck and Clifford W. Collier, Jr. 1995. *Sampling Soils.* West Virginia University Extension Service Fact Sheet. HTTP://WWW.CAF.WVU.EDU/~FORAGE/3201.HTM

The authors wish to thank the following peer reviewers for offering comments to improve the quality and accuracy of the text:

Mary Barbercheck
Department of Entomology
The Pennsylvania State University

Ben Bartlett, DVM
Livestock Extension Educator
Michigan State University

Rebecca S. Bowers
Soil Conservationist
U.S. Department of Agriculture
Natural Resources Conservation Service
West Virginia

Candace S. Burke
State Chair, Pennsylvania Project Grass
Burke's Little Twenty Sheep Farm

Gary C. Burley
East Hill Farms

Robert A. Byers
Frost Entomological Museum
The Pennsylvania State University

Bert Christie
Research Scientist (retired)
Agriculture and Agri-Food Canada

Daryl Clark
Extension Agent
Agriculture and Natural Resources
Ohio State University Extension

Sister Augusta Collins, O.S.B.
Agronomist
The Abbey of Regina Laudis

James B. Cropper
Forage Agronomist
U.S. Department of Agriculture
Natural Resources Conservation Service
North Carolina

Steve DeBroux
Associate Professor of Agronomy and Environmental Science
Delaware Valley College

Sam Dixon
Dairy Manager
Shelburne Farms

Donald C. Fretts
Extension Agent
The Pennsylvania State University

Timothy Fritz
Forage Agronomist
King's AgriSeeds LLC

Jim Gerrish
Consultant, Speaker, Author
American GrazingLands Services LLC

David L. Greene
Principal Agent Emeritus
University of Maryland
College of Agriculture and Natural Resources

James Harrold
Grazing Land Specialist
U.S. Department of Agriculture
Natural Resources Conservation Service
Pennsylvania

Robert Hendershot
State Grassland Conservationist
U.S. Department of Agriculture
Natural Resources Conservation Service
Ohio

James S. Hill
Grassland Specialist (retired)
U.S. Department of Agriculture
Natural Resources Conservation Service
West Virginia

Ronald J. Hoover
On-Farm Research Coordinator
The Pennsylvania State University

Glenn D. Johnson
U.S. Department of Agriculture
Natural Resources Conservation Service
Virginia Polytechnic Institute and State University

Jim Johnston
Pasture Hill Farm

J. Preston Jones
National Program Leader, Agronomy
U.S. Department of Agriculture
Cooperative State Research, Education, and Extension Service

Richard Kersbergen
Extension Professor
University of Maine
Cooperative Extension

Ed Koncle
Farm Seed Sales Manager
Rohrer Seeds

Garry D. Lacefield
Extension Forage Specialist
University of Kentucky

Ralph D. Lentz
Beef Grass Farmer
State Chair,
Grazing Lands Conservation Initiative

R. Clif Little
Assistant Professor
Ohio State University Extension

Larry Lohr
Dairy Farmer
Cold Ridge Farms

John A. Lory
Associate Professor of Extension
University of Missouri

Michael J. Marks
District Conservationist
U.S. Department of Agriculture
Natural Resources Conservation Service
West Virginia

Greg Mullins
Professor, Nutrient Management Specialist
Virginia Polytechnic Institute and State University

Paul R. Peterson
Assistant Professor and Extension Agronomist
University of Minnesota

Dale Ritchey
Soil Scientist
U.S. Department of Agriculture
Agricultural Research Service
West Virginia

Howard Skinner
Plant Ecophysiologist
U.S. Department of Agriculture
Agricultural Research Service
Pennsylvania

S. Ray Smith
Extension Forage Specialist and Associate Professor
Virginia Polytechnic Institute and State University

Kathy J. Soder
Animal Scientist
U.S. Department of Agriculture
Agricultural Research Service
Pennsylvania

Noel Soto
Grazing Lands Conservationist
U.S. Department of Agriculture
Natural Resources Conservation Service
Pennsylvania

Richard Swartzentruber
Farmer Representative for Delaware

Chris D. Teutsch
Assistant Professor
Forage Research and Extension
Virginia Polytechnic Institute and State University

Dan Undersander
Forage Agronomist
University of Wisconsin–Madison

Bill Wayson
Owner
Grace Spring Farm

Donald R. Wild
Conservation Agronomist/
Area Grazing Specialist
U.S. Department of Agriculture
Natural Resources Conservation Service
New York

CHAPTER 1
Plant Morphology and Its Effects on Management

A. Ozzie Abaye, James T. Green, and Edward B. Rayburn

MORPHOLOGICAL CHARACTERISTICS AFFECT FORAGE MANAGEMENT

Forage development, growth, site adaptation, and response to defoliation are closely related to a plant's morphological characteristics. Morphology refers to the outward appearance of plants. Proper grazing and haying management of a perennial forage species is based on the growth characteristics of the species (5). Most grasses have similar morphological characteristics, but modifications from the typical structure allow the various grass species to adapt to specific environmental conditions and management schemes. The morphological characteristics of legumes differ from those of grasses, but morphological differences among the various legume species likewise make them adapted to different management practices. The objective of this chapter is to demonstrate the importance of morphological characteristics in forage management.

MORPHOLOGICAL CHARACTERISTICS OF GRASS AND LEGUME PLANTS

Vegetative Growth

The vegetative stage of the grass plant consists of a collection of shoots or tillers made up of leaf blades and leaf sheaths (figure 1-1). A new tiller arises from a bud located just inside the leaf base

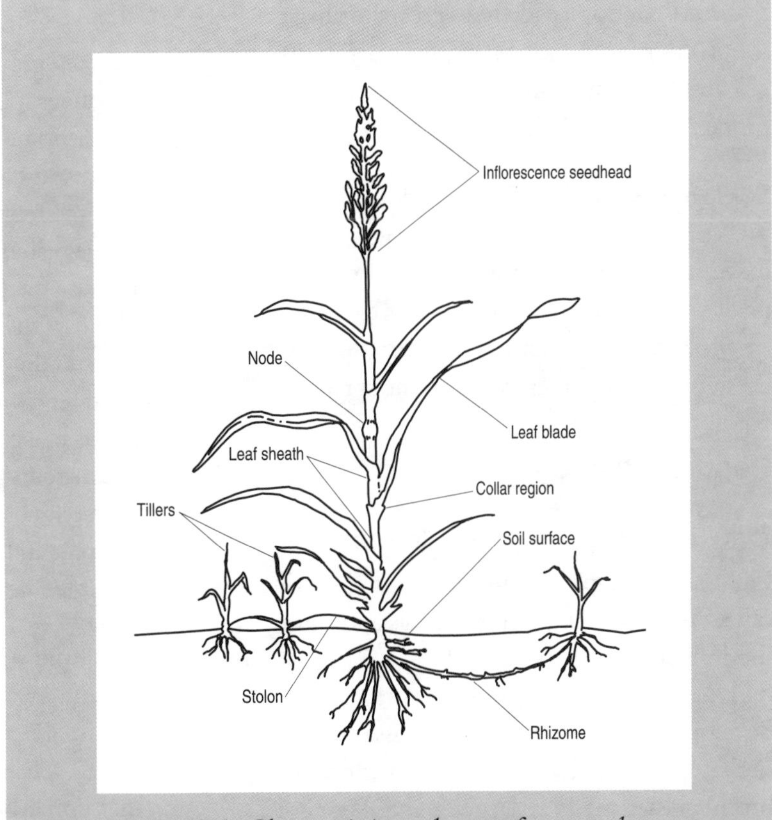

Figure 1-1. Characteristics and parts of a grass plant.

Source: Blaser, R.E. 1986. Managing Forage for Animal Production. *Virginia Agricultural Experiment Station Bulletin 86-7.*

that surrounds the bud. In bunch-type grasses such as timothy and orchardgrass, the new tiller emerges near the base of the parent tiller by breaking through the leaf sheath. In sod-forming grasses the tiller may break through the sheath and form a lateral stem called a rhizome (belowground stem, as in Kentucky bluegrass) or a stolon (aboveground stem, as in bermudagrass). The number and size of tillers per acre determine potential yield.

The leaves arise in succession from a growing point, the location of which relative to the soil surface depends on the forage species. During the year, an aboveground growing point may turn into a seedhead, stem, leaf, node (joint), or dormant bud. Buds are located at the nodes, which are at the tiller bases (basal buds); on the stem (axillary buds); or at the nodes on stolons or rhizomes. Buds provide new growth (tillers/ shoots) during the growing season; fall buds remain dormant to survive the winter and produce next year's tillers.

For legumes such as alfalfa and clover, new growth develops from buds on crowns, stem bases, stolons, or rhizomes. Thus, if the perennial plant is to produce vigorously year to year, it must be allowed to store sufficient energy to survive the winter and start growing in the spring.

Reproductive Growth

Influence of Day Length on Flowering

Light affects plant growth through the production of carbohydrates in photosynthesis and through the influence of the length of the daily light and dark periods (photoperiod) on plant development. Many temperate grasses (e.g., orchardgrass, tall fescue, bluegrasses, timothy, bromegrasses, reed canarygrass, and ryegrasses) require exposure to short days and low temperature in the fall to induce flower bud formation and then warm temperatures and long days in the spring to initiate flowering. These forage species have only one reproductive cycle per year, with flowering occurring in spring. However, the flowering of warm-season grasses, legumes such as alfalfa, and a few cool-season grasses such as prairie bromegrasses is less affected by day length. Day length also affects the vegetative growth of many forage species (6). For example, leaf and stem growth are often upright under long day length, but growth tends to be prostrate under short day length (6). A basic understanding of the influence of day length on flowering and vegetative growth helps determine the type of plant species best adapted to various latitudes.

Managing the Regrowth of Jointing Grasses

Some grass species, such as timothy, reed canarygrass, and smooth bromegrass, can produce stems with elevated growing points or joints without flower production. The elongation of the stem after defoliation elevates the growing point above soil level even though the terminal bud does not become reproductive. Frequent defoliation of these species can eventually reduce their stand and yield because the terminal buds and leaf area for photosynthesis are reduced when carbohydrate levels in the stem base are not sufficient to support rapid regrowth. The production of elevated growing points makes these species less well suited for grazing compared with orchardgrass, tall fescue, or Kentucky bluegrass, which remain vegetative during summer and fall with the growing point buds protected near soil level.

The time when the growing point moves above grazing or cutting height varies among grass species. Timothy and smooth bromegrass have a strong apical dominance (the hormonal control that upper buds exert on the growth of tiller buds in the plant); no basal tillers are apparent until the grass is close to flowering. Cutting or grazing of these grasses should be delayed until the plant enters the reproductive phase. After stem elongation but before flower initiation, the carbohydrate status of the plant is low and new basal tillers have not yet been formed. Thus, cutting or grazing at this point will affect the subsequent regrowth as well as the longevity of these grasses. However, cutting or grazing these grasses before stem elongation removes only leaf blades, so the intact shoot can elongate to produce tillers. Conversely, orchardgrass recovers rapidly when cut at any growth stage because the flowering shoots appear to exert less apical dominance early in the plant's growth than in timothy or smooth bromegrass. Thus, basal buds are available to support growth prior to flowering. However, if an orchardgrass pasture is allowed to go to head (produce seeds), the growth of new tillers will be less than if the seed heads are clipped off, apparently due to apical dominance of the seed heads reducing basal tiller growth.

Location of Growing Points Affects Defoliation Decisions

The location of growing points (meristems) is a key adaptation mechanism for grasses. The location of the growing point on a plant influences how the plant must be managed to balance the relationships among photosynthesis, carbohydrate storage, and growth. The location of the growing point also controls the optimal height of grazing in many cases. Legumes have a terminal bud (meristem) at the end of the stem and axillary buds at each node. Grasses have the same setup, but the stem is short, down inside the leaf sheath, until just prior to flowering. In grasses that have their growing points located near the soil surface (e.g., Kentucky bluegrass, orchardgrass, and ryegrass), expanding leaves on tillers continue to grow after defoliation. These types of grasses are better suited for continuous stocking (grazing) than are grasses whose growing points are elevated by internode elongation (e.g., timothy and smooth bromegrass).

TILLER MANAGEMENT TO INCREASE FORAGE PRODUCTION

Pure Stand

If highly productive perennial grasses are to be persistent year to year, they must be allowed to produce and store enough reserve energy and protein for regrowth and winter survival. A grass tiller is composed of a growing point, leaves, sheath, nodes, stem, and roots. The tiller of a perennial grass can be viewed as an annual or a biennial plant because tillers of forage grasses have a shorter life than the perennial plant has. Therefore, the management of both the existing and potential tillers (those that will grow from inactive buds) is critical to maintaining plant longevity and forage production. Tiller response to defoliation depends on the plant species, the extent of removal of leaf area and growing points, and the amount of reserve carbohydrate and protein available for regrowth. Cool- and warm-season plants such as tall fescue, orchardgrass, alfalfa, birdsfoot trefoil, and lespedeza depend on reserve carbohydrates for regrowth after close cutting or defoliation (8). Therefore, sound management of these grasses should include leaving enough leaf area for carbohydrate production and allowing sufficient time between defoliations to restore carbohydrate reserves.

Mixed Stand (Plus Legumes)

Because plants differ in their morphological and physiological responses to various management strategies, in a mixed species pasture, it is possible to see the dominance of one or two species over the others at any given time. The species selected for a mixture should be morphologically compatible and have relatively similar maturity dates. In a grazing situation, the animals' selection of one species or plant part in preference to another depends on relative palatability, acceptability, and overall availability of pasture species or parts.

Legumes and grasses vary in their growth pattern as well as their response to defoliation. Figures 1-2 and 1-3 show the effect of grazing height on bluegrass-white clover and alfalfa-orchardgrass mixtures, respectively. When a bluegrass and white clover mixture is grazed to 1-inch height, the regrowth of bluegrass is slower than white clover, thus favoring the dominance of white clover in this mixed sward (figure 1-2). At 1-inch grazing height, the small and horizontal white clover leaves can escape grazing while the semierect bluegrass leaves are easily grazed off (1). However, at 2-inch grazing height, bluegrass regrowth will be faster, which will shade the white clover, hence causing a shift in botanical composition favoring the dominance of bluegrass over white clover. The cutting height of 2 inches or higher favors bluegrass over white clover because new growth of bluegrass begins rapidly from the apex leaves; this causes a decrease in light intensity reaching the developing white clover leaves arising at soil level from stolons (figure 1-2).

Similarly, when alfalfa and orchardgrass are grown in a mixture, close grazing or cutting to 1.5-inch height (figure 1-3) depresses growth of orchardgrass, thus reducing light competition and favoring alfalfa dominance (1). After grazing or cutting, new growth of alfalfa comes from the stem bases and crowns near or below the soil surface; new growth of orchardgrass comes from leaf area remaining. Therefore, leaf area remaining

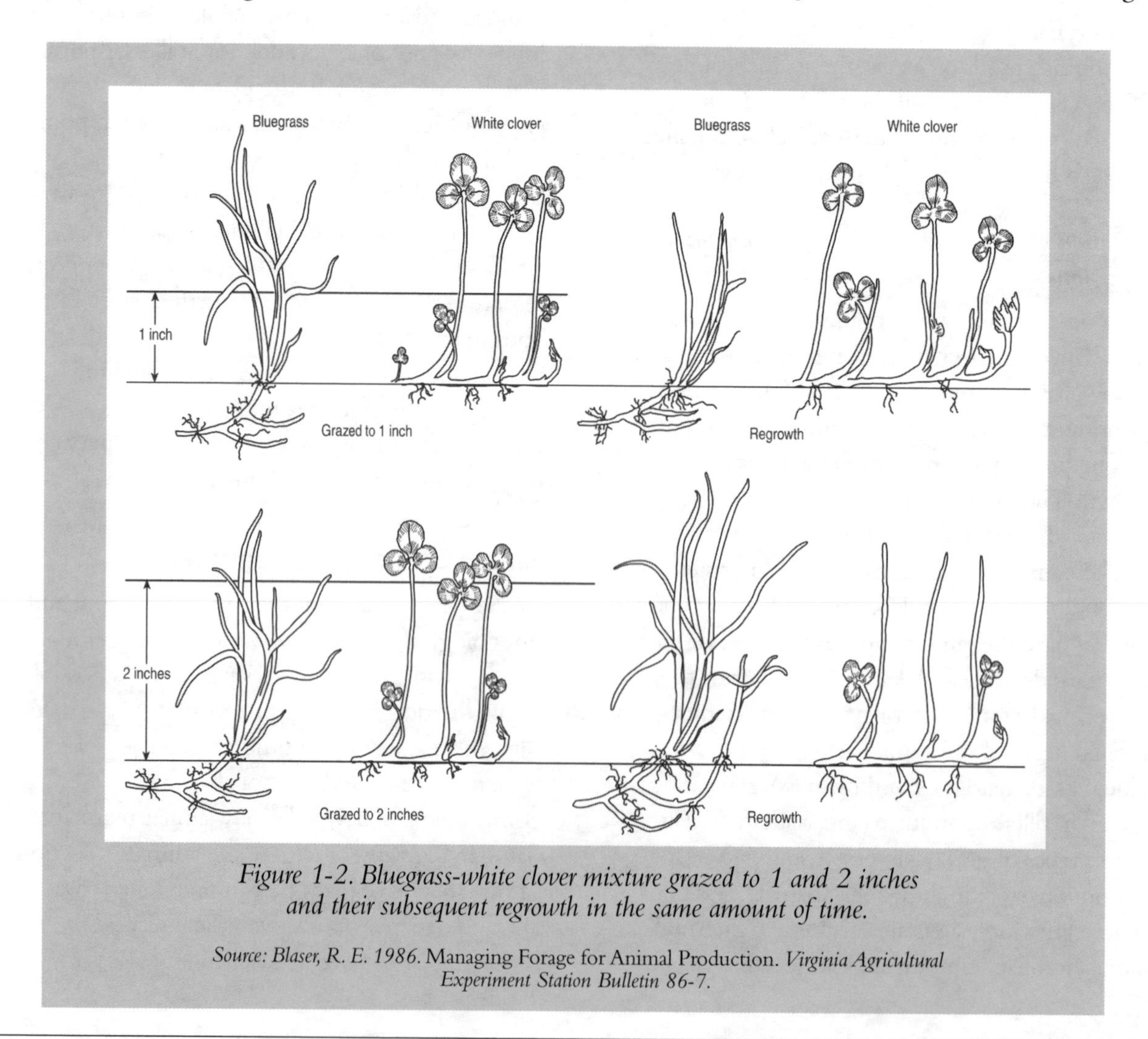

Figure 1-2. Bluegrass-white clover mixture grazed to 1 and 2 inches and their subsequent regrowth in the same amount of time.

Source: Blaser, R. E. 1986. Managing Forage for Animal Production. *Virginia Agricultural Experiment Station Bulletin 86-7.*

after cutting or grazing is less critical for alfalfa than orchardgrass. Regardless of similarity in morphological characteristics of mixed stands, the manager must make a management decision to favor the forage component that is less competitive in order to maintain a reasonable balance between the species in the mixture.

THE ROLE OF SHADING IN THE SURVIVAL OF DEVELOPING TILLERS

The immediate effect of leaf shading is lowered photosynthesis, and consequently, reduced production of carbohydrates. Carbohydrate production by the leaf declines further due to changes in physiological and morphological characteristics of leaves in reduced light (4). Shading also affects dry matter (everything in the plant except for the water) distribution within the plant and plant development. Shading increases leaf length, internode length, and plant height, but reduces leaf width and thickness, rate of leaf initiation, and root/shoot ratios, as well as tillering in grasses and branching in legumes (3, 9). The number of tillers in grasses and branches in legumes is reduced in part by the lower rate of leaf appearance, which provides fewer leaf buds from which new tillers can develop, and in part because fewer of these buds develop tillers (4). Thus, the number of tillers, rather than tiller size, is most influenced by shading. Because the nitrogen

Figure 1-3. Alfalfa-orchardgrass mixtures grazed to 1.5 and 3 inches and their subsequent regrowth in the same amount of time.

Source: Blaser, R. E. 1986. Managing Forage for Animal Production. *Virginia Agricultural Experiment Station Bulletin 86-7.*

fixation rate of legumes is directly related to carbohydrate supply, shading reduces both nodulation and nitrogen fixation due to reduction in energy receipt (4).

Managing Forage for Acceptable Leaf Area Index

Given favorable growing conditions (moisture and temperature), maximum growth of temperate grasses and legumes occurs when leaves accumulate to intercept about 90% of the incoming light. Grasses and legumes require different leaf area indexes (LAI—total area of all leaves on the plant to the area of ground covered by the plant) to capture 90% of the light. To do this, clovers, alfalfa, and most grasses require leaf area index units of 3–5, 5–6, and 7–11, respectively (figure 1-4). Leaf area in pastures is proportional to plant height, so plant height can be used in place of LAI (on the x axis, figure 1-4) to estimate light interception.

When no other factor is limiting growth, LAI can be used to make a management decision to control the interval between defoliations. The accumulation of leaf area in excess of that needed to intercept 90% of the light can result in a decline in pasture growth rate. This is due to shading and leaf aging reducing the photosynthetic ability of

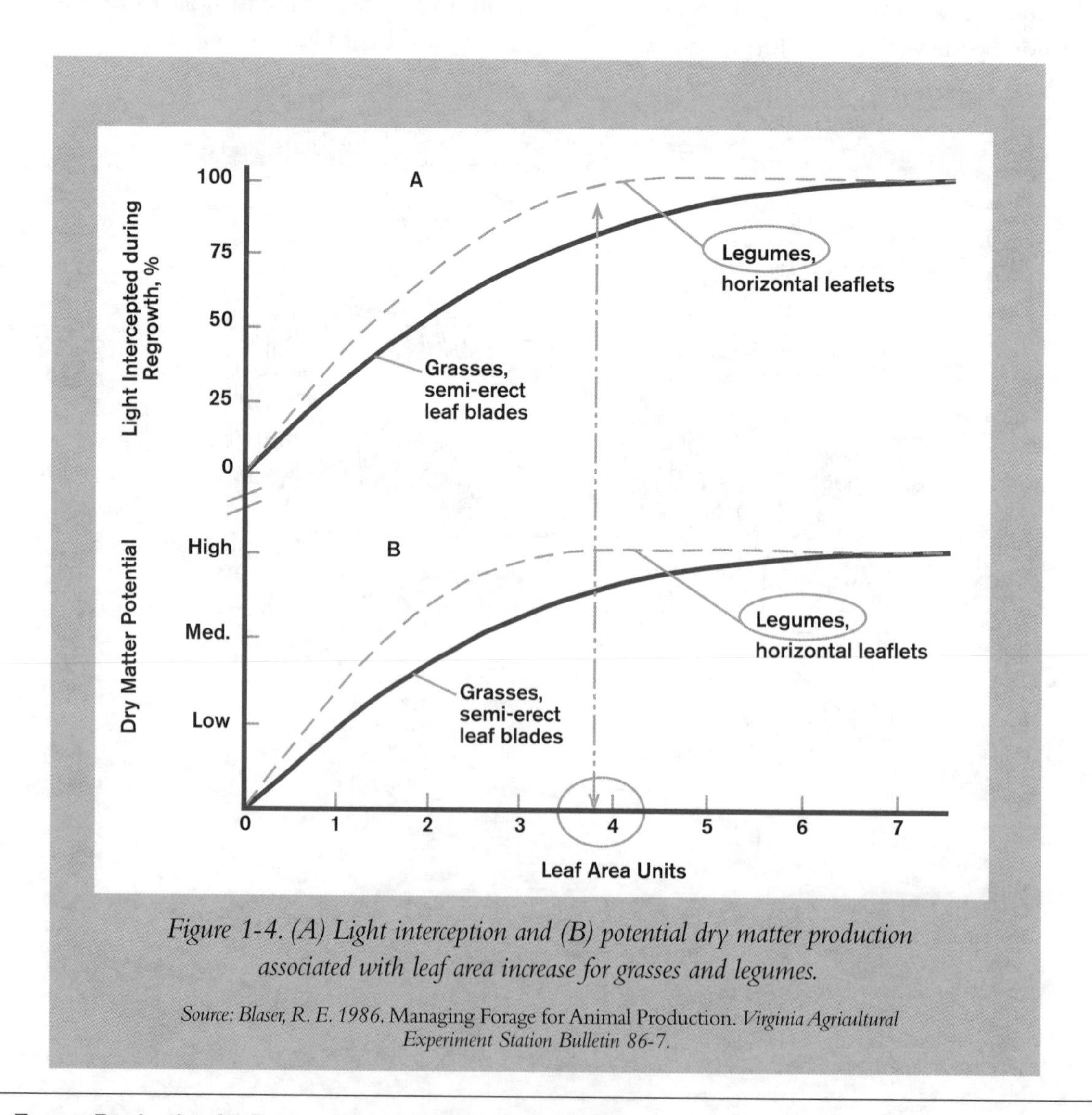

Figure 1-4. (A) Light interception and (B) potential dry matter production associated with leaf area increase for grasses and legumes.

Source: Blaser, R. E. 1986. Managing Forage for Animal Production. *Virginia Agricultural Experiment Station Bulletin 86-7.*

individual leaves and leads to leaf death. Since the growth rate of leaves is determined by both environmental and management factors, it is necessary to vary grazing intervals based on the time of year and weather patterns. For example, when the sun is at a lower angle in the sky in the fall, a shorter plant height is needed to intercept 90% or more of the sunlight (7) (figure 1-5). Depending on the management goal, this may mean that pastures can be grazed lower in the fall.

Vertical Versus Horizontal Growth

The arrangement and orientation of leaf growth within the canopy affects the amount of light intercepted and the amount of carbohydrates produced. Leaves that have a horizontal orientation (e.g., white clover, crimson clover) absorb a greater amount of light at a relatively low LAI as compared to plants with vertical leaves (e.g., tall fescue), which allow light to penetrate deeper into the crop canopy.

The management implication of vertical versus horizontal plant growth habit is very important. Short-growing plants with horizontal growth habit can be grazed or cut frequently to avoid overlapping of leaves, which results in leaf shading and rapid leaf aging. Legumes such as alfalfa, with a tall, vertical growth habit, allow more light to penetrate to the lower canopy, resulting in higher

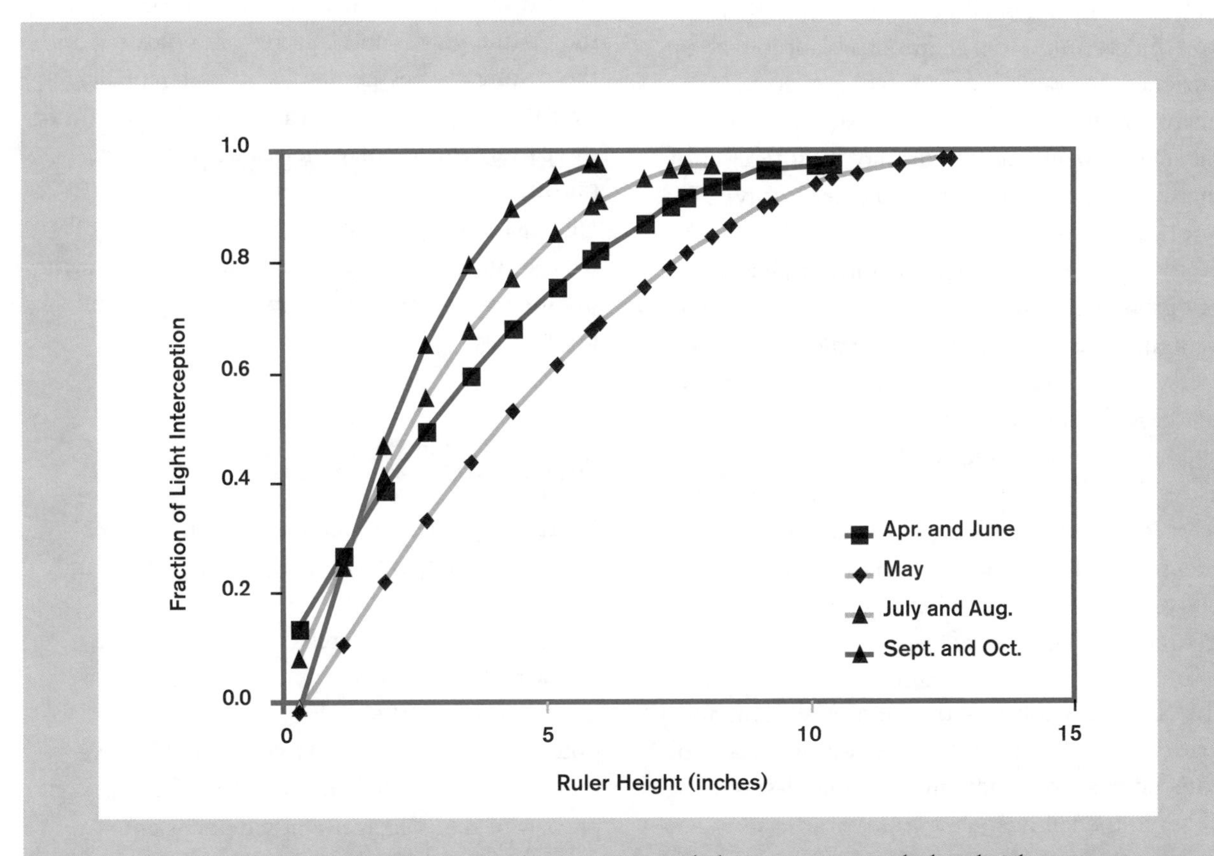

Figure 1-5. The relationship between percent light interception and plant height.

Source: Rayburn, E. B., and S. B. Rayburn. 1998. A standardized plate meter for estimating pasture mass in on-farm research trials. Agronomy Journal *90: 238–241.*

growth potential and yield. However, because most of the leaf area is removed by cutting or grazing or dies due to aging or shading, regrowth of these plants is mostly dependent on stored carbohydrates. Short-growing plants with leaves close to the soil surface (e.g., white clover, Kentucky bluegrass) are tolerant of close grazing because the leaves are not totally defoliated with close grazing. These plants are also less dependent on stored energy for recovery because they can produce photosynthetically active leaves near the soil surface.

ROOT SYSTEMS

The roots play a primary role in the plant's anchorage and water and mineral uptake. Root growth determines the ability of a plant to take up nutrients and water. Root growth is affected by intrinsic root/shoot partitioning patterns and the soil environment (e.g., soil fertility, moisture, aeration, drainage, structure, and bulk density). Root growth is determined by the plant's actively photosynthesizing leaf area. Defoliation reduces root growth, depending on the cutting height and the frequency of defoliation. The amount of reserved carbohydrate prior to harvest also affects root growth and recovery (1). If defoliation is frequent and removes most of the top growth (leaves), available carbohydrates become a limiting factor for root respiration. Consequently, root extension and water and nutrient uptake from the soil are all substantially reduced. Root death can occur at high rates of defoliation.

The effect of defoliation on the root/shoot ratio varies by the seasonal cycle of root growth; as cool temperatures reduce top growth of cool-season grasses in the fall, a larger amount of photosynthate goes to root growth (2).

Forage species differ in morphology, depth, and distribution of roots. For example, smooth bromegrass and Kentucky bluegrass, which appear different morphologically aboveground, have a similar root mass. However, a higher proportion of smooth bromegrass roots is distributed deeper in the soil profile than that of Kentucky bluegrass. This gives smooth bromegrass a competitive advantage when mixed with Kentucky bluegrass under drought conditions. Species with deep taproots, such as alfalfa, compete successfully against shallow-rooted grasses when lack of soil moisture is the limiting factor. However, when summer rainfall occurs as light frequent showers, shallow-rooted plants may have a competitive advantage once water reserves deeper in the soil are used up.

The botanical composition of pasture can be altered when weeds that are deeper rooted than the pasture grasses and legumes are not eaten and the pasture species are defoliated preferentially over the weeds. The growth of the root systems of the pasture species may then be restricted if grazing pressure is not controlled and weeds encroach. An understanding of the response of roots to grazing or cutting is essential in achieving the full benefits of legumes and grasses for animal production.

SUMMARY

Morphological characteristics of plants dictate forage development, growth, site adaptation, and response to defoliation. The tall, erect-growing forage species such as orchardgrass, tall fescue, alfalfa, and red clovers are best used for hay or silage because of their relatively high yield potential, which justifies harvest costs. However, these forages are also suitable for grazing if a proper grazing management scheme is employed to allow for sufficient rest periods and residual leaf area for carbohydrate production after grazing. Short- and prostrate-growing plants such as bluegrass and white clover are best suited for grazing and may be stocked continuously. The low and prostrate growth habit of these plants allows them to produce leaves close to the soil surface, thus escaping continuous removal of leaves by the grazing animals.

Generally, defoliation that is too frequent or too intensive reduces forage yield. Regrowth after defoliation depends on the location and activity of growing points, leaf area left for photosynthesis, carbohydrate and protein reserves in the plant, and the rate of recovery of root growth for nutrient and water uptake. The relative impact of defoliation on growth is determined by environmental aspects such as temperature and moisture, in addition to the plant's adaptive characteristics to defoliation frequency and intensity. Proper grazing and/or cutting management of temperate grasses and legumes in pure or mixed stands depends on a few general factors: leaf area (for photosynthesis), stored carbohydrate (for regrowth or winter survival), the location of the growing point, the origin of new growth tillers (e.g., stem base, stolon, or rhizome), and finally, the root. If high production and stand longevity of grasses and legumes are to be maintained, the manager must integrate each one of the above factors into the daily management schemes.

Chapter 2
Ecology of Plant Communities in Forage-Livestock Systems

E. Ann Clark, Heather Karsten, William M. Murphy, and Benjamin F. Tracy

DEFINITION AND SCOPE

Ecology is the study of interactions among system components. In this chapter, we focus on the management of plant community–based interactions, including those among plant and animal species, soil and climate, and fencing and watering systems, to enhance the profitability of grass-based livestock systems.

The plant community consists of the sown and unsown species that make up the sward. However, what is sown as a single, uniform plant community typically becomes a mosaic of different communities because species differ in their adaptation to growing conditions, which, in turn, differ within and among paddocks (plate 1). Particularly in perennial pastures, plant communities come to mirror the environmental heterogeneity in which they grow, as soil texture (e.g., the relative distribution of sand, silt, and clay particles), drainage, or grazing history effects accumulate over the years, selecting for or against individual species.

Succession in sown perennial pastures includes species diversification—an increase in species number—and stratification—self-sorting to produce a mosaic of diverse plant communities at different locations (2). Is resisting these natural tendencies in order to maintain a uniform sward a sensible goal? The cost of frequent reseeding or other management to maintain uniformity must be weighed against the value of the improved animal performance from a uniform sward. Before a producer can realize a net profit from a pasture reseeding, the newly sown mixture has to produce more than the original pasture, for enough years, to pay off the cost of the reseeding (table 2-1).

Plate 1. Mosaic of plant communities in a sown pasture, with patches of orchardgrass or bromegrass separated by a matrix of white clover/ Kentucky bluegrass.

Table 2- 1. Does it pay to reseed?

Establishment cost of reseeding	Amortization years	Yearly payback (principal + 10% interest)[a]	% Increase in net return required to break even
$200/ac	3	$80	160
	5	$52	100
$100/ac	3	$40	80
	5	$26	50

[a] *Assumes net return prior to reseeding was $50/acre. For example, to pay off $200/ac in establishment costs, plus 10% interest (total of $240), in 3 years requires that the newly sown pasture net an additional $80 on top of the original $50 (e.g., total of $130/ac from land originally netting $50/ac), and it must do this for 3 years in a row, just to break even.*

Assuming the same level of management, do new seedings pay for themselves in a realistic time frame? On-farm surveys in the United Kingdom show that sown species' contribution to pasture yield declines exponentially with time. After 5 and 10 years, sown grass species accounted for an estimated 55% and 40% of sward yield, respectively, although the rate of loss was slower at high rates of nitrogen (N) fertilizer. Sward performance would presumably decline with the loss in sown species.

Less costly approaches to sward improvement, such as frost-seeding of legumes coupled with planned disturbance or shifting management to favor desired species without reseeding, may enhance the economic feasibility of sward improvement. An alternative approach that will be discussed below is to learn how to use sward diversity as a strength to buffer and sustain pasture performance throughout the grazing season.

It could be argued that the single factor most discouraging producer interest in management-intensive grazing (MiG) is the belief that unlike grain crops, pasture performance is uncontrollable, at the whim of the weather and the land base, and hence, too unpredictable to be worth the effort. A central goal of this chapter is to demonstrate that management of plant community–based interactions is a tool, just like seed or fertilizer, enabling creation and maintenance of pasture communities to serve specific functions. A complementary goal is to show how fencing and grazing management can be employed to use each type of plant community to its strength.

USING INTERACTIONS TO CREATE MANAGEABLE ASSETS

Pasture species interact with each other and with many factors known to vary within and among farms. Plants interact with soil factors (chapter 3), climatic variation, pests and diseases, the demands and stresses imposed by various types of livestock, fertility (chapter 4), and grazing management. Every plant species responds to environmental stresses, but the degree of response differs among species, as discussed in chapter 1 and below.

Known differences in adaptation among species may be used to influence species' persistence and shift composition within the pasture to meet specific needs (table 2-2).

For example, the site characteristics and desired species composition to support a spring turnout pasture will be different from those for midsummer grazing or fall grazing/winter feeding (table 2-3, p. 15). No single species or simple mixture of species can meet all of these diverse needs, year after year. A diversity of pasture soils, sward types, and management regimes enables producers to better accommodate seasonal variability in pasture growth and buffer against the vagaries of the weather.

The yield stabilizing effect of a more complex mixture is shown in a Pennsylvania study comparing small plot performance of eight mixtures ranging in complexity from 1 to 15 commonly planted pasture species (e.g., white clover, tall fescue, and orchardgrass) (adapted from (4)). Figure 2-1 (p. 16) shows how variation in yield (shown as the coefficient of variation or CV) over the growing season responds to (A) an index of plant diversity (Shannon's index) (3) and (B) the standing crop. The negative slope of the lines shows that as the diversity (and standing crop) of the plots increased, yield variability

Table 2-2. Using plant community–based interactions to enhance pasture performance.

Interaction	Example/opportunity	Management solution
Plant: plant	When sown as a complex mixture, reed canarygrass may never appear because it establishes slowly and is suppressed by more aggressive species. Conversely, the fraction of the sward that will come from perennial ryegrass is well out of proportion to its share of a seed mixture, because it establishes so readily (plate 2).	Appropriate mixture composition, adjusted relative seeding rates, and post-seeding grazing management can compensate for species-specific differences in establishment rate.
Plant: soil	In an uneven field, a complex mixture may sort itself into timothy, Kentucky bluegrass, and white clover in the low spots, with bromegrass and alfalfa more dominant on well drained knolls.	If you can't beat them, join them! Fence to separate ridges from gullies and use each to its strength (see pp. 19–23) (plate 3).
Plant: climate	Alfalfa overwintering is favored by leaving a tall stubble to trap snow and resist icing. Yet both orchardgrass and perennial ryegrass are subject to snow molds when left too tall going into the winter (plate 4, p. 14).	Consider these kinds of incompatibilities when designing a mixture. Also, once frosted, lightly graze off excess herbage to reduce organic residue and minimize risk.
Plant: grazing management	In a mixture of tall and short species, hay management will favor tall, sparsely tillering species, leaving bare ground vulnerable to airborne weed establishment between the plants. Conversely, frequent, intensive defoliation will favor short, dense tillering that occupies space and discourages weed encroachment, but yields less (plate 5, p. 14), particularly under droughty summer conditions.	Withholding spring grazing to allow a first hay cut every 2 or 3 years in a season-long grazing rotation can help to maintain both types of species in the sward, while reducing weed encroachment. Overseeding white clover will help to fill in open space and discourage weeds.
Plant: fertility management	Difficulties in retaining legumes in a sward may reflect a deficiency of phosphorus and/or potassium, on which legumes are particularly dependent. Legumes are also penalized at moderately low pH levels that are still tolerated by many grass species.	Consider both nutrient and pH factors when analyzing legume persistence problems. Overapplication of N, as fertilizer or manure, can favor grasses over legumes.

Plate 2. Fall establishment success from a midseason seeding of perennial ryegrass (right) and orchardgrass (left).

Plate 3. Fencing on the contour to separate dissimilar plant communities created by topography, soil type and depth, and exposure.

Plate 4. Snow mold on orchardgrass.

Plate 5. A mixture of tall and short pasture species.

Table 2-3. Different grazing applications need different sites and sward types.

Application	Site features	Sward features
Spring turnout	Well drained, preferably on a south-facing slope, with suitable windbreak	Early, sod-forming species able to recover from treading and compaction (plate 6), such as Kentucky bluegrass and white clover
Midsummer grazing	Low-lying, heavy soil, with good access to soil moisture	Tolerance to imperfect drainage, with midseason vigorous growth, such as birdsfoot trefoil, red clover, and reed canarygrass
	Well drained land, suitable for machinery	Hay crop aftermath grazing, suitable for bringing into the rotation to balance supply and demand in summer, such as alfalfa-based mixtures
Extended fall grazing and/or winter feeding	Well drained, protected from wind, preferably in need of nutrient input	Late fall-growing, sod-forming species tolerant to grazing under wet conditions; an old, well established sod, such as tall fescue

Plate 6. Treading damage on wet soil.

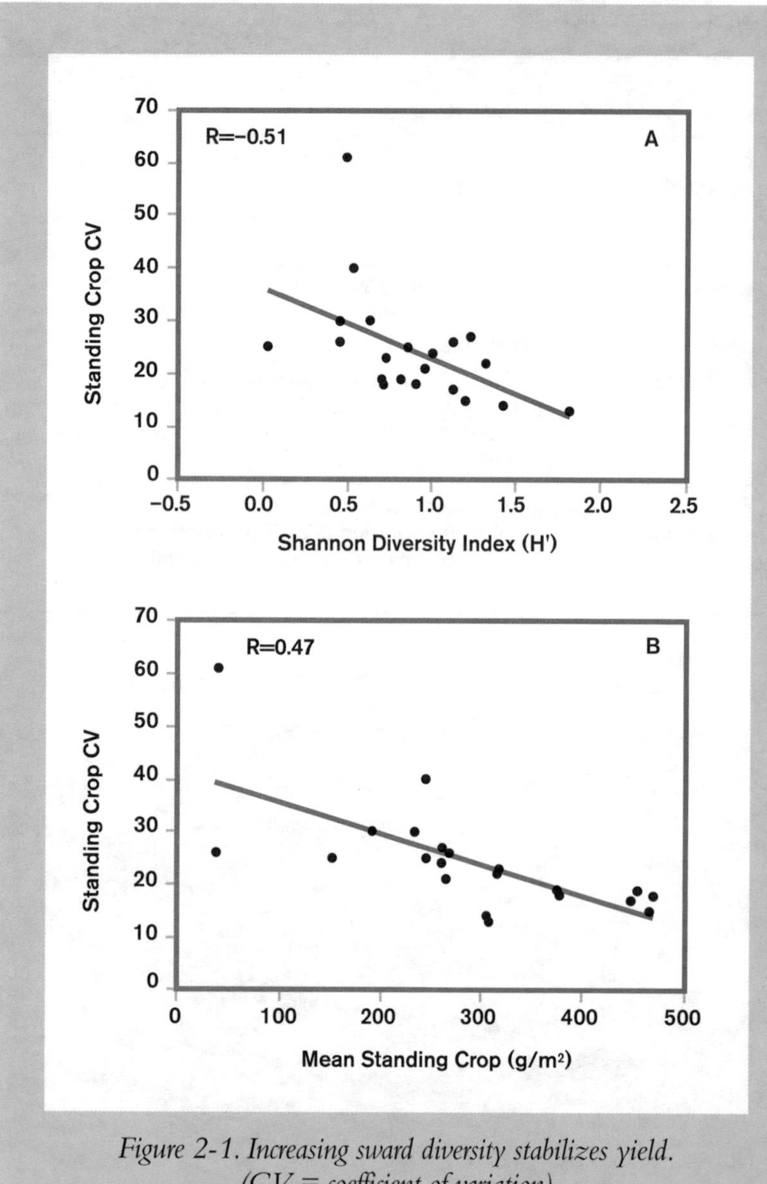

Figure 2-1. Increasing sward diversity stabilizes yield. (CV = coefficient of variation)

Source: Adapted from Tracy, B. F., and M.A. Sanderson. 2004. Productivity and stability relationships in clipped pasture communities of varying species composition. Crop Science *44:2180–2186.*

fluctuations. For example, drought-tolerant species such as orchardgrass may be more advantageous during dry than wet times. Under wet conditions, however, species that are better adapted to higher soil moisture conditions, such as timothy, may compensate. The end result is more consistent production over the growing season, and across years differing in weather.

Interestingly, not only did the complexity of mixtures affect biomass stability, but it also influenced the degree of weed invasion into plots. The most complex mixtures (>4 species) averaged about 80% less weed biomass than more simple mixtures. This suggests that greater sown diversity repressed weed invasion by using available resources (e.g., sunlight and soil water) more completely, and thus did not give weeds an opportunity to take hold. Maintaining high diversity in pastures may be a way to reduce weed problems.

decreased. In effect, yield was more stable as diversity increased.

This result occurred mainly because complex mixtures (>3 species) usually contained species that each responded slightly differently to climate

PLANT FUNCTIONAL TRAITS

Sustaining pasture productivity within and among years requires a diverse array of species, but not just any group of species. It is necessary to select species that not only fit well with each other and that are adapted to the growing environment, but that also serve all the key functions you require from the pasture (e.g., table 2-3, p. 15). For example, a producer who ensures that his or her mixture includes species that are able to (i) fix N, (ii) build a sod, and (iii) grow during a drought has prioritized these three functional traits. A producer who manages a mixed sward to retain both cool- and warm-season species is relying on the staggered growth cycles of these two functional traits to stabilize production. In an environment where perennials are slow to establish, including some annuals in a perennial seeding exploits the complementary growth patterns of these two functional traits (establishment vigor) to reduce time out of production.

The particular functional traits that you choose to include in your pasture will depend entirely on what is important to you. If the pasture is intended to be used for just a few years, speed of establishment is a key function. Species that establish rapidly, such as perennial ryegrass, timothy, orchardgrass, and alfalfa, would confer this preferred functional trait. Conversely, speed of establishment may be less of an issue for a longer term pasture, which could therefore include slower establishing species such as reed canarygrass, bromegrass, and Kentucky bluegrass. If the pasture is intended to receive manure from a confinement facility, species that quickly and efficiently take up N, such as orchardgrass and reed canarygrass, would confer a key functional trait. The legume functional trait would be of lesser importance here than in most pastures.

A partial list of some common types of functional traits is provided below. Functional traits are not mutually exclusive. A given species can express more than one trait, depending on the attributes of interest. Not all functional traits are necessary or appropriate for every pasture. The point is to choose diversity strategically, with specific aims in mind.

Legumes

Legumes fix N symbiotically, and thus require less soil N than nonlegumes. Most are valued for productivity during hotter, drier weather, when cool-season grasses slow down. They are typically higher in protein and more rapidly digestible than grasses, although much of the protein is rumen-degradable rather than rumen-bypass. Although all can fix N, legumes differ in height, drought tolerance, and adaptation to grazing. *Alfalfa, birdsfoot trefoil,* and *red, Kura,* and *alsike clovers* are taller and more competitive for light. In contrast, *white clover* is a short, shade-sensitive species with horizontal stems called stolons that spread laterally. It thrives under closer and more frequent grazing than the other legumes.

Rhizomatous, Dense Sod-Forming, Cool-Season Perennial Grasses

Rhizomatous, dense sod-forming, cool-season perennial grasses fill in open spaces, reduce weed invasion, and protect the soil from compaction, degradation, and erosion. *Kentucky bluegrass* and *tall fescue* share the same basic morphology, but differ in other respects. For example, tall fescue is more drought-tolerant than Kentucky bluegrass. It also grows much later in the fall and retains its nutritional value better under the snow, leading to its reputation for winter grazing. However, bluegrass is much more palatable and nutritious, and was a valued component of the famous fattening pastures in the Midlands of the United Kingdom. Some cultivars of tall fescue can be endophyte-infested.

Rhizomatous, Open Sod-Forming, Cool-Season Perennial Grasses

Rhizomatous, open sod-forming, cool-season perennial grasses include tall, jointing grasses such as *reed canarygrass* and *bromegrass,* which are valued for high seasonal productivity. However, much of their yield comes from stem tissue. Although a weed species, *quackgrass* is closely related to the cultivated wheatgrasses and is valued for both productivity and persistence. Although all are tall, these species differ in other attributes. Bromegrass is very large-seeded and establishes slowly; reed canarygrass is slower still. Both bromegrass and quackgrass are more tolerant of the herbicide glyphosate than other grasses, which can allow them to proliferate in no-till seedings.

Bunch-Type, Cool-Season Perennial Grasses

Bunch-type, cool-season perennial grasses are valued for spring and fall growth. Both *perennial ryegrass* and *timothy* are unproductive in the heat and drought of summer, but *orchardgrass, meadow foxtail,* and *meadow brome* are less drought-sensitive. Perennial ryegrass, which is available in both forage-type cultivars and shorter turf-type cultivars, often appears to be a sod-forming species because of dense tillering. Perennial ryegrass has been bred to be highly palatable and nutritious and highly responsive to N, although some cultivars can be endophyte-infested.

Warm-Season Annuals

Warm-season annuals are sown when hot, dry periods are a regular occurrence. Small acreages can also be sown as a precaution, to provide emergency feed where drought is a possibility. Examples would include *sorghum* x *sudangrass hybrids* and *sudangrass*—both of which carry a risk of prussic acid (also known as hydrocyanic acid or HCN) poisoning—as well as *corn* and *pearlmillet.*

Warm-Season Perennials

Warm-season perennials are popular where summers are typically hot and dry, helping to stabilize production during the summer slump. However, their requirement for warmer soils means a late start in the spring, which can leave a window for weed encroachment if sown in monoculture. *Switchgrass* is one example.

Annual or Perennial Forbs

Annual or perennial forbs (broadleaf species that are not legumes) produce palatable, highly digestible, and nutritious forage. The annual brassicas, such as *turnips, kale,* and *fodder rape,* are sown in spring for summer grazing or in summer to extend the fall grazing season. *Chicory,* a deep-rooting perennial, is valued for summer grazing.

Species Productive on Imperfectly Drained Soils

Species productive on imperfectly drained soils include *reed canarygrass, meadow foxtail, birdsfoot trefoil,* and *red* and *alsike clovers. White clover* can also occur on wet soils, typically due to lateral spreading during the growing season.

Understanding the particular adaptations of individual species and their specific contributions to your overall pasture plan will help determine an appropriate mixture for each field or grazing unit. Anticipating how each species will respond to grazing management, fertility, and other plants in the sward will also guide your efforts to shift or maintain desired mixture components.

SOURCES OF ENVIRONMENTAL VARIABILITY IN MANAGED GRASSLANDS

Environmental variation can be caused by natural factors as well as management decisions. Both large-scale or landscape factors and small-scale or within-paddock factors can affect the growing environment. Because of species-specific differences in adaptation, sward composition often comes to vary within and among paddocks. Variable sward composition within a given paddock is a recipe for nonuniform grazing, reducing harvest efficiency and profitability. Because grazing preference varies with plant species and plant maturity, less favored or unpalatable species (i.e., increasers) can displace desired species (i.e., decreasers) when species composition differs too greatly within a paddock.

For these reasons, environmental—and hence, sward—variability has traditionally been viewed as a liability. To a large extent, graziers have been persuaded by the same logic that motivates hay producers: to simplify management, maintain sward uniformity by frequent replanting. However, as the Pennsylvania study showed (pp. 12–16), sward variability can be a strength rather than a liability. We propose that:

- *Large-scale* environmental differences (e.g., soil type, slope, and aspect) that create diverse plant communities should be addressed by fencing.[1] Dissimilar pasture communities, whether sown or occurring through natural processes, should be separated into different paddocks, allowing each plant community to be used to its strength.
- *Small-scale* environmental variability, which can favor different species in the space of a few feet, can be accommodated by sowing or managing for complex mixtures of three or more species to discourage weed encroachment and sustain productivity. Sowing a wider range of desirable species acts as a genetic buffer, allowing plants to occupy diverse niches and exclude weeds that could otherwise encroach.

Topography

Topography is a predictable source of large-scale variability in a pasture. Species that thrive on the drier uplands are likely to be different from those on the face of a slope or the lower, wetter land. Thus, to allow each zone to be used to its strength, sloping land can be fenced into upland, slope face, and lowland grazing zones (plate 7).

Plate 7. Fencing to create different topographic grazing units.

[1] *Where fencing is precluded because of distance or other obstacles, herding can achieve the same results.*

Permanent fences can be placed on key lines—where ridges change to hillsides and where hillsides change to valleys or terraces. Temporary fence can be used to further subdivide the zones into manageable paddocks.

Two Ontario studies (Cases 1 and 2 below) on long-term pastures illustrate how slope position, grazing management, and sward age interact to stratify sward composition within the same pasture. Several useful principles are illustrated by each study.

Case 1

A 30-year-old commercial pasture in Bruce County, the heart of the Ontario beef industry, was monitored for 3 years following subdivision into three large paddocks (Clark, unpublished). The producer created the paddocks by fencing down, rather than across, the slope. Over 3 years, this pasture supported 1–1.25 steers per acre for about 150 days of grazing, producing just over 300 pounds per acre of gain per year. The following summarizes the results of that study.

1. *Increasers and Decreasers.* Meadow fescue, a less palatable species, accounted for 40–60% of available herbage in the pasture (figure 2-2). Uncontrolled grazing had given stock the latitude to bypass less preferred species and allowed species such as the fescues to expand to occupy an ever larger share of the land base. Higher stocking rates and controlled grazing to oblige more uniform grazing could have helped to bring back the desired species at the expense of the less preferred species. However, this would have been feasible only if fencing had been placed on the contour, rather than down the slope face, to separate dissimilar swards and concentrate grazing appropriately.
2. *Dry vs. Wet.* Orchardgrass, a dryland-adapted species, was more prominent on the drier uplands ("top"), but timothy and Kentucky bluegrass showed a stronger adaptation to the wetter and heavier soils at the bottom of the hill ("bot") (figure 2-2). Upland and lowland species grow and respond to management differently. Thus, to allow each zone and its plant community to be managed to its strength, fencing would have to run across rather than down the slope.

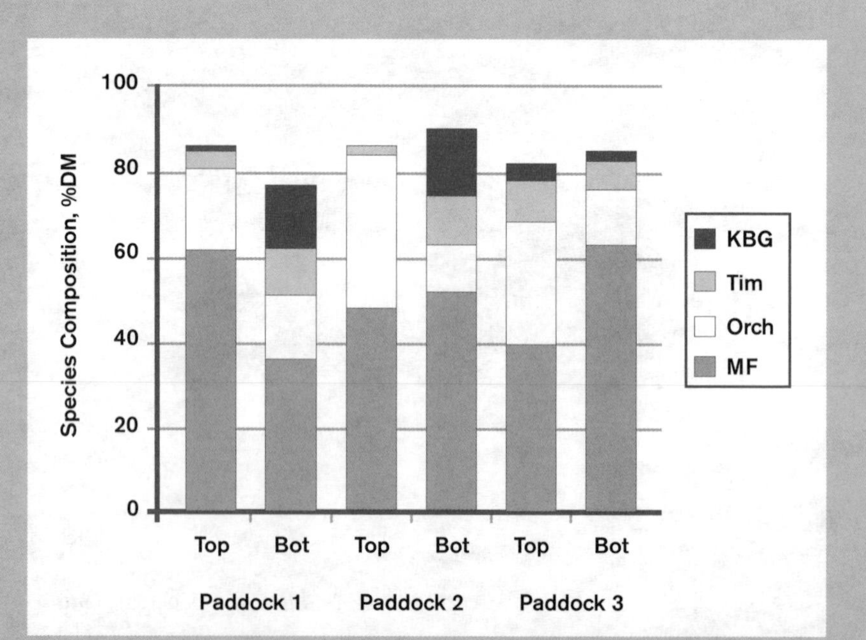

Figure 2-2. Differences in predominant species composition among paddocks and position in paddock in a 30-year-old commercial beef pasture in Ontario (Clark, unpublished) (KBG=Kentucky bluegrass, Tim=timothy, Orch=orchardgrass, MF=meadow fescue).

3. *Mixture Complexity.* Four species accounted for about 80% of the sward in both positions in all three paddocks (figure 2-2). The balance was made up of other species, including smooth brome, quackgrass, and narrow leaf plantain in upland positions, and redtop, white clover, and tall buttercup in the bottom land. Each of these secondary species contributed up to 5–7% of sward yield in specific locations, but was absent in other parts of the same paddock. Thus, 10 species (including three weeds) made up the bulk of the sward; an additional 14 other weed species and sown legumes were present in small quantities. Sufficient environmental heterogeneity existed in this 30-year-old pasture to support 24 species, although the contribution of each species varied greatly across the landscape. This study illustrates the breadth of variability that becomes evident over years in commercial pastures and supports the need for fencing on a macro-scale and for sowing more complex mixtures to fill available microscale niches to maintain composition and discourage weed encroachment.

Is the degree of diversification and stratification seen in this commercial beef pasture a reflection of the class of land and level of management applied, or do the same trends occur on better land, under more intensive management?

Case 2

The 50-acre Northfork research pasture in Ontario was originally sown in 1985 to a complex mixture of six grasses and three legumes, with some legume frost seedings in later years. For comparison purposes, the Northfork facility typically supported the equivalent of 1–1.5 steers per acre for a 150-day grazing season, plus shortkeep stock, producing the equivalent of 500–800 pounds gained per acre. After 10 years of MiG, species composition on this moderately sloping, Class I, tile-drained land was assessed (plate 8) (adapted from (2)).

Plate 8. Aerial view of the Northfork grazing research facility near Elora, Ontario.

1. *Increasers and Decreasers.* As before, a single species, Kentucky bluegrass, expanded over time to occupy a dominant position, although for a different reason than in Case 1 above (figure 2-3). Previous research showed that the short stature and sod-forming character of Kentucky bluegrass made it a superior competitor under the intensive grazing regimes under investigation at Northfork (figure 2-3). Thus, the phenomenon of increasers and decreasers can be propelled by differences in adaptation as well as palatability.
2. *Dry vs. Wet.* As before, although sown uniformly, species stratified over 10 years of grazing along zones of adaptation (figure 2-3). Orchardgrass and white clover persisted on the better drained upland (rep. 1) and sloping land (reps. 2–3), but were replaced by meadow foxtail and quackgrass in the wetter lowland (rep. 4).
3. *Mixture Complexity.* Six species (including two weed species) accounted for about 90% of sward yield (figure 2-3), although the relative contribution of each one varied from nil to 60% within and among reps. Bromegrass, meadow brome, birdsfoot trefoil, alfalfa, and weed species occurred sporadically, although each was prominent at specific sampling locations because of small-scale environmental heterogeneity. The minimal contribution of weed species to the sward, even after 10 years of grazing, supports the value of complex mixtures in providing sufficient genetic variability to accommodate small-scale heterogeneity and exclude weeds.

Thus, although younger, more tightly managed, and on productive, tile-drained land, the Northfork sward demonstrated the same kind of stratification evident in the commercial pasture. Diversification and stratification are to be expected even on well managed pastures.

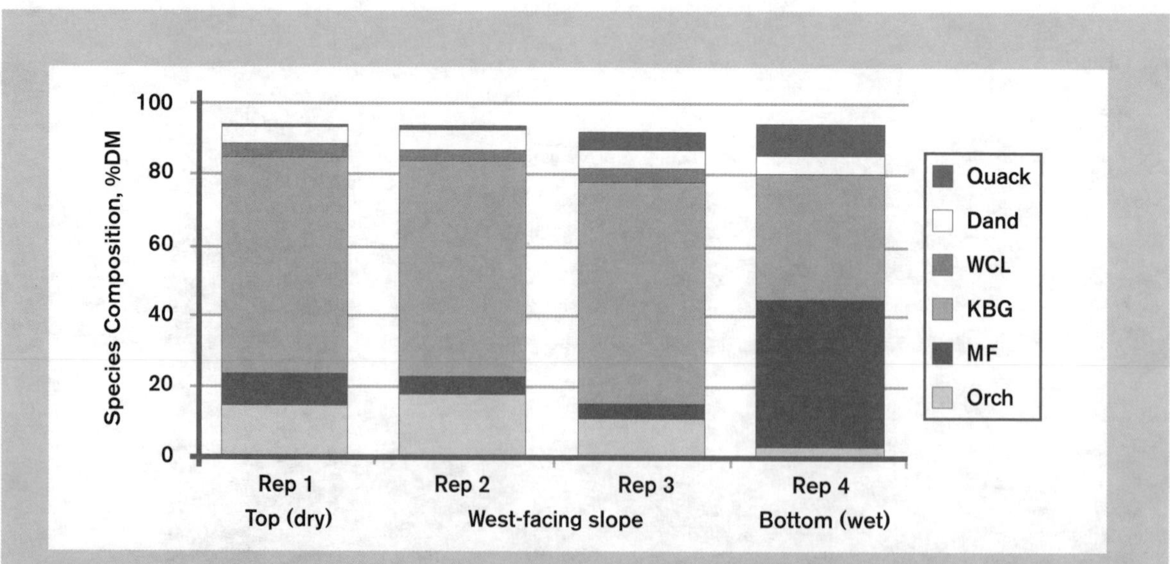

Figure 2-3. Differences in species composition with slope position in a 10-year-old research pasture in Ontario. (Quack=quackgrass, Dand=dandelion, WCL=white clover, KBG=Kentucky bluegrass, MF=meadow fescue, Orch=orchardgrass).

Source: Adapted from Clark, E. Ann. 2001. Diversity and stability in humid temperate pastures. pp. 103–118, In: P. G. Tow and A. Lazenby (ed.). Competition and Succession in Pastures. *CAB International Publishing, New York, N.Y.*

Large-scale stratification means that management of even moderately sloping land is improved by fencing into distinct topographic zones, allowing each zone to be used independently to support grazing. Contemporary flexible fencing minimizes the need for bracing, square corners, and straight lines, allowing fence to follow the contour of the land and soil and better reflect the layout of the various ecological zones (plate 7, p. 19).

The small-scale niche diversity present on Class I tile-drained land of the Northfork pasture is just a fraction of that in a more typical pasture. As revealed in the Pennsylvania study discussed on pp. 12–16, complex mixtures of sown species are needed to occupy the niches present in each zone, exclude undesirable weeds, and stabilize yield.

Soil Texture

Prior to the advent of precision agriculture, an agricultural field was managed as a single soil type. In reality, a given field is typically a mosaic of intermingled soil types (plate 9, p. 24) with varying abilities to support plant growth. Soil textural differences related to topography can be amplified by overgrazing or by continuous row cropping, which enriches lowlands at the expense of the uplands. However, texture can also differ across a landscape, independent of current topography. Soil texture differences will affect drainage and water-holding capacity, and hence, species composition, growth, and ultimately, soil structure. These differences are a liability, allowing overgrazing of preferred species as well as undergrazing of less preferred species, unless controlled with strategically placed fencing.

Weeds

Not all weeds are undesirable, particularly if the goal is profitability rather than productivity or appearance. Quackgrass and especially dandelions are as palatable and nutritious as any sown species at the same stage of development. Nonetheless, a high density of dandelions would be undesirable because they produce low yield. Weed encroachment is not so much a problem in itself, as a symptom of a larger problem. Weeds grow because they are better adapted to whatever you are doing–grazing, fertility management, etc.–than whatever you planted. Concentrating on killing the weeds is dysfunctional, because they will just come back anyway. Focus instead on adjusting management to favor the desired species, while disfavoring the toxic, noxious, or unpalatable species.

A dense, solid sod with few bare spots is the most effective means of excluding weeds, whether from the wind or from the soil seed bank. Both overgrazing and undergrazing can create gaps in the sod that encourage weeds, although for different reasons. Overgrazing (or camping, or winter feeding, see pp. 27–29) decimates the plant community, exposing the soil and providing an opening for weed growth. Undergrazing or delayed haying allows a tall, lanky canopy to develop, shading the base of the plant and suppressing the growth of the small tillers upon which regrowth depends. When the top growth is finally removed, only dead tiller bases and bare soil remain–an invitation to weeds. So remove the excess, through haying or grazing or topclipping, in a timely fashion, both to exclude weeds and to create the capacity for speedy regrowth.

Plate 9. Typical soil survey map of a portion of northern Wellington County, Ontario. Each shade corresponds to a different soil type.

Source: Compiled, drawn, and published by the Soil Research Institute, Research Branch, Canada Department of Agriculture, Ottawa, 1962, from base maps supplied by the Department of Mines and Technical Surveys, Ottawa.

Textural zones large enough to warrant fencing should be managed separately. Talk to your county extension agent or local USDA-Natural Resources Conservation Service staff if you need help distinguishing areas of different soil types. Grazing practices, including seasonality and intensity of use, can then reflect the productive capabilities of the different zones. Sandier soils can support either grazing early and late in the season or winter feeding, when moisture excess rather than deficiency is of concern. Susceptibility to compaction and soil disruption (plate 10, p. 25) outweigh the potential for plant growth on more clayey soils. The strength of clayey soil zones comes in midseason grazing, when moisture deficit limits growth elsewhere on the farm. Species best adapted to sandy soils include smooth and meadow brome, orchardgrass, tall fescue, and alfalfa. Clayey, imperfectly drained soils could support timothy, reed canarygrass, Kentucky bluegrass, birdsfoot trefoil, and red, white, and alsike clovers.

Fencing to separate different soil textural zones can refine pasture utilization, enhance harvest efficiency, and avoid lasting damage to the pasture. Shifting stock into sandier, dry paddocks during protracted rainy periods can minimize compaction and treading damage that would occur on clayey soils. Provision of fenced paddocks on heavy clay soil to support grazing during protracted drought minimizes evaporative moisture loss, replacement of desirable species by weeds, and vulnerability to erosion that could occur on sandier soils. Problem avoidance, by design, is arguably a cheaper and easier solution than problem solving after the fact.

Grazing Systems

Topography and soil texture and drainage are natural factors that have been shown to affect species composition. However, species composition can also be changed by management, including the design and implementation of a grazing system. The good news is that management can be tailored to shift species composition to meet identified needs. The bad news is that species shifts can also occur unintentionally, through inappropriate management.

Grazing animals can cause rapid and large changes in plant productivity and botanical composition of the pasture sward. In addition to physical effects exerted through treading and voiding (see pp. 27–29), livestock can also affect pasture performance through defoliation, as moderated by the grazing preferences and behavior of the particular class of livestock.

Defoliation

Defoliation is probably the most important effect that grazing animals have on pasture. Plant species differ in tolerance to the timing and intensity of defoliation, in their growth rhythms during the season (see chapter 1), and in their attractiveness or palatability to animals. As a result, grazing pressure, climatic conditions following grazing, and species-specific differences in adaptation interact to shift species composition during the season. For example, intensive spring grazing promotes clovers, while spring application of N tends to favor N-responsive grasses and depress species diversity.

Broadly speaking, species may be categorized as pasture-adapted (e.g., Kentucky bluegrass, perennial ryegrass, and white clover) or hay-adapted (e.g., jointing grasses, such as bromegrass, timothy, and reed canarygrass, as well as alfalfa and birdsfoot trefoil). This distinction is based on physiological and morphological adaptations that protect and buffer carbohydrate reserves and growing points against grazing pressure. However, there are many species that fall somewhere in between, including orchardgrass and tall fescue.

Carbohydrate reserves act as a buffer to maintain plant growth and metabolism during periods of shortfall, such as following defoliation or during drought stress. Removing the photosynthetic surface (the leaves) of a plant through grazing reduces or even stops root growth because the plant must readjust the balance between root and shoot functions. When leaves are removed, roots typically die back to regain the appropriate balance

Plate 10. Soil disruption and erosion around winter feeding sites.

(see root pulsing, next column). To reestablish new leaf area following grazing, through extension of immature leaves or unfolding of new leaves, grasses depend on reserves for 2–7 days, compared to 21 days or longer for alfalfa. It follows that species that maintain more basal leaf area following grazing, and are thus able to continue photosynthesis, are less dependent upon stored reserves and are also more tolerant of frequent, intensive defoliation.

Low temperatures and/or dry conditions slow the regrowth of defoliated plants, which prolongs dependence on carbohydrate reserves. Uncontrolled continuous grazing, by repeatedly removing leaves and reducing reserves during and after adverse weather, decreases survival and regrowth of stressed plants. In northerly latitudes, development of cold hardiness in species such as alfalfa may be lessened either by defoliation during the critical fall harvest period (6 weeks prior to the first killing frost) or by excessive grazing during the season.

Defoliation can also benefit plant growth and the larger environment. Appropriate levels of defoliation cause pasture plants to tiller and branch more profusely, producing a tighter sod that protects the soil and safeguards against weed encroachment. With white clover, more branching and stolon development, together with less shading, encourages flowering and seed set. This tendency may be very important in maintaining red and white clovers in intensively grazed pastures.

Deep-rooted plants also cycle nutrients vertically from lower depths in the soil profile. Because grazing animals retain only a small fraction of the nutrients they ingest, the remainder is redistributed onto the pasture through voiding, where it becomes available for uptake by other plants. In this way, defoliation and digestion by grazing animals promote rapid cycling and transfer of nutrients to the pasture. Conversely, uneaten plant tissues slow nutrient cycling because nutrients are immobilized and unavailable until the plant material breaks down through weathering or biological decomposition. Controlled high stocking densities reduce the amount of uneaten plant material, and through hoof action help to break down uneaten material, promoting decomposition and quickening nutrient cycling.

Root pulsing (periodic growth and die-back of roots) through defoliation and regrowth is another underappreciated benefit of managed grasslands. Root dieback following defoliation adds organic matter and carbon to the soil. Root growth under managed grazing effectively channels more carbon from the atmosphere to the soil, thereby helping to lower atmospheric carbon dioxide and reduce the risk of global warming. Increasing the total land area in grassland and improving grazing management could help to reduce atmospheric carbon dioxide.

Selective Grazing

Animals don't graze uniformly, like a mowing machine. Animals select leaves over stems, young tissues over old, and preferred species over unpreferred species, if allowed to do so. Uneven grazing not only reduces utilization efficiency, but also changes species composition by favoring less palatable species while disadvantaging preferred species (see pp. 20–21).

With proper fencing and a high stocking density, animals can be obliged to consume less palatable species. The goal may be to graze down an overmature sward to stimulate regrowth in support of high-demand animals, to reverse encroachment by less palatable species, such as the fescues, or to control weeds. However, intake will decline, reducing weight gain. Thus, when grazing specifically for sward management, as to shift species composition, it is common to use clean-up animals, such as dry cows or ewes.

Herd Impact Effects

Livestock concentration affects soil properties, manure distribution, and hence, sward composition. Soil disturbance and degradation caused by treading and trampling is commonly observed in camping and other high-traffic areas. The term "camping" refers to the congregation of livestock on hill tops, under trees, or around water tanks or feeders. However, by definition, treading and trampling also occur to varying degrees throughout a pasture. The key to sward longevity is to regulate the timing and degree of treading within the tolerance limits of both the sward and underlying soil. The vast grasslands that evolved under the intense grazing pressure of migrating ungulates are testament to the resilience of grasses to treading, so long as sufficient time is allowed for recovery.

All livestock can cause treading damage, particularly when soil is moist and grazing is unmanaged. A single animal makes 8,000–10,000 hoof prints each day. A single cattle hoof print covers 14 square inches, so a total of 0.02 acres per day is stepped on by a single animal.

In Wales, a clay loam soil that had been intensively grazed by dairy cattle for 26 years was subjected to mechanical slitting to a depth of 6 inches in the spring. In that year herbage yields doubled, and uptake of N, phosphorus (P), and potassium (K) from the soil also increased. The beneficial effect of mechanical slitting was attributed to improved aeration, which effectively broke open a zone of impermeable soil located 3.9–4.7 inches deep in the soil that had produced a perched water table (i.e., water could not penetrate below the impermeable layer). The depth of compaction within the profile varies with soil type, and tends to be deeper on lighter soils. If you doubt the compactive effect that grazing can have, try to take soil samples from grazed versus hayed land!

Natural remedies for compacted soil include the beneficial effects of earthworms. Within a week of removing stock from a rotationally grazed paddock, hoof marks may be seen to contain earthworm castings. Freezing and thawing are other processes that help to break up compacted soil.

Treading indirectly affects plants by compacting the soil (increasing bulk density) and by encouraging puddling (reducing air space), especially when soils are moist and fine-textured (clayey). Damaging the soil increases resistance to root growth, reducing water and nutrient uptake as well as N fixation. Compacted soil retards water infiltration and promotes runoff, leaving less water to support plant growth while also encouraging erosion. Compared to ungrazed pasture, moderate and heavy grazing can reduce infiltration rates by 25 and 50%, respectively, with essentially no infiltration occurring in the most heavily trafficked areas.

Plant species differ in tolerance to treading. Perennial ryegrass, Kentucky bluegrass, tall fescue, white clover, and Kura clover are most resistant, and orchardgrass, timothy, birdsfoot trefoil, red clover, and alfalfa are least resistant. Species composition can shift to favor the more resistant species as stocking density increases. Nonetheless, because plants, animals, and soils evolved together, treading has always been part of the selection pressure. Grazing and probably the hoof action of treading can be beneficial in breaking up clumped mats of dead tissue that would otherwise shade basal tillers and reduce growth.

The plant community in heavily trafficked camping areas is often quite distinguishable. Where vegetation is worn down to expose bare soil, characteristic weeds can appear. Weeds may come from the soil seed bank or from the windblown seed (seed rain). Either way, bare soil provides the niche for weeds to establish. Canada thistle is a particularly common invader of disturbed sites, often forming a patch with a well defined edge corresponding to the limits of the camping area. The presence of annual and biennial weeds, which are seldom seen in a well managed perennial pasture, is an indicator of mismanagement, such as allowing bare soil to appear.

Unless carefully chosen and managed, winter feeding areas can be problematic, leading to soil erosion and degradation. Livestock concentration also means manure accumulation, flies, and risk of both overland flow and contamination of water supplies by nutrients and pathogens. Cattle defecate 11–12 times a day and urinate 8–11 times a day, and each mature animal typically produces 50 pounds of manure a day. At 10 inches in diameter, enough cowpies are produced by one cow to cover 1,260 square feet during the grazing season! Cattle void approximately 80% of the nutrients they consume, amounting to 38, 8, and 8 pounds of N, P, and K, respectively, per animal during the grazing season. Thus, allowing manure to concentrate in any one area can significantly reduce soil fertility elsewhere in the pasture. It is to your benefit to manage winter grazing, just as summer grazing, through fencing, placement of water, and timely rotation, to diffuse and redistribute manure.

Manure concentration from winter feeding can also be employed as a strength if feeding is strategically positioned to supply nutrients to low fertility zones. Selection of winter feeding sites provides a useful tool for moving considerable nutrients within and among farm fields without the expense of manure hauling and spreading. Shifting winter feeding sites from year to year may also perform a useful function in promoting sward diversification and rejuvenation.

Unavoidable concentration, camping, and feeding areas should be designed to minimize adverse environmental impacts. Permanent watering or feeding sites may be underlain with a cement or gravel pad to reduce mud and allow accumulated manure to be collected and redistributed to the pasture. Innovative producers in Renfrew County, Ontario, devised a cost-effective raft of rough, delimbed cedar trees lashed together with wire upon which the water tank at a community pasture rested.

Although harrowing should not be necessary to break up manure pies and redistribute manure, it may be advisable for areas of livestock concentration. On a Vermont dairy employing leader-follower grazing on a Kentucky bluegrass-white clover sward, no benefit was obtained from harrowing (Murphy, unpublished). New Zealand grass consultant Vaughan Jones places great emphasis on earthworms as agents of rapid manure decomposition and removal, even to the point of applying calcium specifically to promote earthworm activity, independent of pH. Areas of plant rejection around manure pies—a common sight on many pastures—are not visible on the farms of his clients. Dung beetles, soil invertebrates, and other decomposers play a key role in rapidly reintegrating manure nutrients and organic matter into pasture soils. Due to both the movement of nutrients from grazed to camping areas and the potential for environmental damage, excessive levels of manure nutrients should be redistributed back to the land from areas of unavoidable or intentional livestock concentration.

Short-duration rotational grazing will minimize exposure of any given site to trampling and disruption. Dairy grazing for 12-hour increments, through six rotational grazing cycles a year, amounts to 3 days of use per year, leaving 362 days for rest and recovery. High-density, short-duration grazing simulates the selection pressures under which grasses evolved. Alternatively, disruption can be applied intentionally to expose legume seed in the soil seed bank and rejuvenate a sward. Charlie Opitz, a well known Wisconsin dairyman, intentionally disturbs his permanent grass swards every few years by applying heavy grazing pressure under wet conditions. Mudding up the paddocks has the effect of exposing seed and stimulating germination of red clover, a biennial that would otherwise have to be periodically reseeded.

If damage unavoidably recurs, it may be advisable to fence off the affected area—temporarily or permanently—for rest or reseeding. The high-traffic areas may be sown to a tough, sod-forming species, such as tall fescue or Kentucky bluegrass. Emerging perennial weeds such as Canada thistle can be removed manually or spot-sprayed to discourage spread.

Shifting Sward Composition

Both livestock behavior and species-specific differences in plant adaptation to grazing give producers the means for intentionally shifting and altering species composition, regardless of initial seeding rate and mixture. Controlled continuous grazing or intensive rotational grazing with a relatively short rest interval will suppress taller, more erect-growing species, commonly resulting in a Kentucky bluegrass-white clover–based sward. Conversely, a regime of infrequent harvest and lengthy regrowth intervals will suppress these species in favor of taller, coarser species such as alfalfa and timothy. This information can be used to sustain swards intended specifically for pasture or for hay usage. However, it can also guide grazing decisions when it is desirable to maintain both hay- and pasture-adapted species in the same sward.

For example, pasture-adapted species tend to be more drought-sensitive than hay-adapted species, reducing their utility for midseason grazing. Where midsummer drought is likely, it may be desirable to maintain some pasture land with species chosen for midseason growth, such as birdsfoot trefoil, to take the pressure off of the rest of the pasture. Alternatively, it is possible to retain hay-adapted cool-season species, such as alfalfa, red clover, and the bromegrasses, as well as warm-season grasses in a pasture sward. To do this requires a defoliation regime that sustains species with differing adaptations in the same sward at the same time. This is easier than it sounds, because herbage species are both perennial and pliable. Composition of a perennial sward reflects both its immediate and its long-term history. What is in a sward now reflects the life experience of the sward, in terms of both management and climate. A pasture sward is a dynamic living community. Small differences in rate of tiller production and death among neighboring plants can swiftly change species composition. Vulnerability to change is an overlooked but highly useful characteristic of a pasture.

So, how do you retain species of differing adaptation? Changes in entry and exit heights and regrowth durations within and among years can maintain differently adapted species in the same sward. As shown in a Pennsylvania study, even moderate differences in entry and exit height materially influence species persistence and performance (1). A mixed sward of 30% orchardgrass, 29% Kentucky bluegrass, 17% quackgrass, 9% dandelion, 7% legumes, and 8% other species was subjected to two height-based grazing regimes: entry and exit at 10.6 and 2.8 inches, respectively (tall) vs. entry and exit at 8.2 and 2 inches, respectively (short). Over 2.3 years, herbage harvested averaged 51% more at each grazing event in the tall regime, with orchardgrass accounting for most of the difference in harvest (figure 2-4, p. 30). Tall-growing legumes, such as alfalfa and red clover, also benefited from the higher entry and exit heights. Measurable changes in herbage growth were stimulated within a 2-year time frame.

Changing management among years can be particularly effective in preventing dominance by

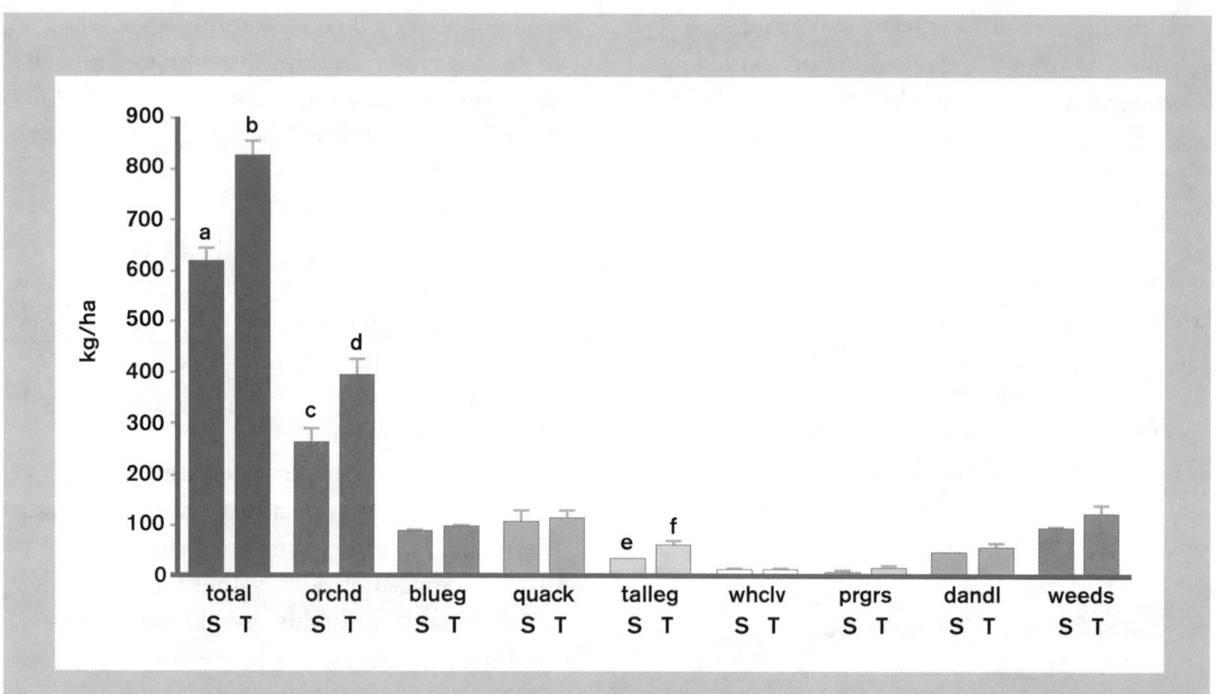

Figure 2-4. Species contribution to available forage under short and tall grazing regimes (S=short grazing regime, T=tall grazing regime; orchd=orchardgrass, blueg=bluegrass, quack=quackgrass, talleg=tall legumes, whclv=white clover, prgrs=perennial ryegrass, dandl=dandelion).

Source: Adapted from Carlassare, M., and H. D. Karsten. 2002. Species contribution to seasonal productivity of a mixed pasture under two sward grazing height regimes. Agronomy Journal *94: 840–850.*

either hay- or pasture-adapted species. One innovative Ontario sheep producer was worried about the excessive growth of white clover that was promoted by season-long rotational grazing on his farm. He split the farm into two halves, of which one was rotationally grazed season-long, and the other was withheld from grazing for a first hay cut, with rotational grazing of the aftermath. By swapping management regimes every 2 years, he was able to rebalance the sward, favoring the coarse grasses at the expense of the white clover.

Similarly, an Ontario study subjected three grass species to intensive rotational grazing by cattle, with entry and exit heights of 7.9–9.8 inches and 2–2.4 inches, respectively (Clark, unpublished). Three years of this management regime greatly favored white clover, at the expense of each grass species (table 2-4). Withholding spring grazing in Year 4 to allow a first hay cut rejuvenated each grass, although bromegrass had been sufficiently debilitated to prevent meaningful recovery.

In this example, partial sward rebalancing was induced simply by replacing two cycles of spring grazing with a single hay cut. Thus, the built-in memory of perennial swards, coupled with the dynamism of tiller creation and death in differently adapted species, gives graziers the ability to shift species composition at will to serve their particular needs. Haying or topclipping most paddocks once a year can be helpful in balancing sward composition and controlling weeds.

Table 2-4. Sward rebalancing, measured as percent contribution of grass vs. white clover to mixture yield,[a] caused by withholding spring grazing in Year 4.

	Year 2		Year 3		Year 4	
	% Grass	% Clover	% Grass	% Clover	% Grass	% Clover
Orchardgrass	40	46	14	72	22	58
Bromegrass	24	61	5	75	7	69
Reed canarygrass	27	44	21	65	40	41

a Measured in midseason, as the mean of three cultivars per grass species, in four-, two-, and three-species mixtures; the balance of yield was made up of other sown species and weeds.

SUMMARY

Plant species differ in adaptation to each other and to their growing environment, as well as in the functions they perform within an overall pasture plan. The complex interactions that determine species persistence and, ultimately, sward composition, are conditioned by forces operating at the landscape level, as well as those active on a much smaller scale. Species-specific differences in adaptation to the growing environment are the root cause for the diversification and stratification that are so characteristic of perennial pasture swards. Frequent reseeding to maintain a uniform sward of one or two species, in opposition to these natural tendencies, shortens the interval over which reseeding costs can be amortized. An ecologically based alternative is to map pastureland into resource units based on natural features of the land, allowing the grazier to tailor grazing to the unique strengths of each unit.

Topography and soil texture were presented as examples of natural resource endowments that are known to affect the growing environment and, hence, sward composition, in predictable ways. Fencing on the contour to separate the dissimilar swards that evolve at various slope positions or on various soil textures allows each sward type to contribute in its area of strength. Sowing or managing for more complex mixtures within a given paddock supplies a sufficient range of genetic variability to occupy available niches, exclude undesirable weeds, and sustain productivity.

In addition to coping with natural sources of environmental diversity, graziers can create and harness plant community diversity—independent of seeding—by strategic management. As one example, graziers can use species-specific differences in response to grazing pressure and herd impact to favor hay- or pasture-adapted species, or maintain both plant types, at will. Management is a tool, just like seed, fertilizer, or drainage, to control the plant community and help achieve desired ends. The cost-effectiveness of working with what you have, instead of regularly replacing it with a new sward, should be considered by profit-minded graziers.

CHAPTER 3
Soils, Soil Fertility, and Fertilizers

Edward B. Rayburn and Thomas J. Basden

INTRODUCTION

Cost-effective fertilization management on livestock farms involves effective cycling of nutrients harvested during grazing and conserving stored feeds. Nutrients needed beyond those recycled can then be provided by legume management and purchased fertilizers and lime.

SOILS

Soils hold water and nutrients needed by plants for growth. The physical, chemical, and biological characteristics of a soil determine which plants will survive on the site and how well they will grow. Soils are classified based on their physical and chemical characteristics. These characteristics are determined by the parent material on which the soil developed, the climate under which it developed, the plant and animal life living on and in it as it developed, and the slope position and drainage characteristics where it developed.

The U.S. Department of Agriculture's Natural Resources Conservation Service (NRCS) has surveyed most counties to identify the various soil types across the landscape. The results are published in soil surveys, which include descriptions of the soil types and maps showing where they occur. These soil survey maps are available at NRCS, conservation district, and extension offices. When an NRCS or conservation district technician develops a farm conservation plan, a map of the soils on the farm is part of the plan.

It is important to know the drainage, rooting depth, and fertility characteristics of the soils in each field across a farm. Having this information allows the manager to select the forage species and varieties most adapted to the field.

Some soils have natural physical problems such as poor drainage. Poor drainage may be caused by the presence of a hardpan or impervious layer that restricts internal drainage. Poor drainage may also occur at the bottom of a slope where runoff from above collects. Soil drainage influences which forage species will survive on a field as well as the microbiology and chemistry occurring in the soil; the latter two factors determine the effectiveness of nutrient cycling.

Droughty soils can result from (i) coarse soil texture (sands, gravels, and stoniness), which holds little water per foot of rooting depth, (ii) low organic matter in a fine-textured (clayey) soil, resulting in poor soil structure and reduced water infiltration and water-holding capacity, or (iii) shallow rooting depths caused by a hardpan or bedrock.

Prior management can adversely affect soil physical properties and crop productivity. Soil compaction can occur when crop tillage reduces soil organic matter or heavy machinery causes compaction. Soil erosion results in the loss of soil water-holding capacity by removing the upper soil horizon, which is higher in organic matter. Soils that have experienced serious erosion are often lower yielding than those with less or no erosion.

Soil pH and fertility are determined by soil type and prior management. Where soil pH or fertility is below that needed for optimal plant growth, these factors can be corrected by lime and fertilizer applications. Soil testing determines the soil's pH and fertility status (see pp. 34–39). Plant response and local economics determine if fertilizer and lime applications are economically justified (see pp. 39–50).

THE REALISTIC YIELD EXPECTATION OF A CROP ON A SOIL

The realistic yield expectation (RYE) of a crop grown on a soil is determined largely by the amount of plant-available water held in the rooting zone (figure 3-1) (8, 9, 10). This is determined by the soil's texture, organic matter content, and plant rooting depth. Plant rooting depth is determined by the soil and by the rooting characteristics of the species grown.

Soil surveys include a table listing the RYE for selected crops and pasture when grown under good management on each soil type in the county (table 3-1, p. 34). Note the difference for pasture yield expressed as animal unit grazing days (AUD—a 1,000-pound animal grazing for 1 day) per acre for the Albright silt loam (200 AUD) compared to the 12–25% slope Dekalb channery loam (105 AUD). A copy of a soil yield table for your county should be available from the local county extension agent or NRCS office.

The RYE values of soils in a county are based on local experience and research and are a reliable first estimate of potential crop yields under good management. However, with improved plant varieties and management, it is possible to exceed

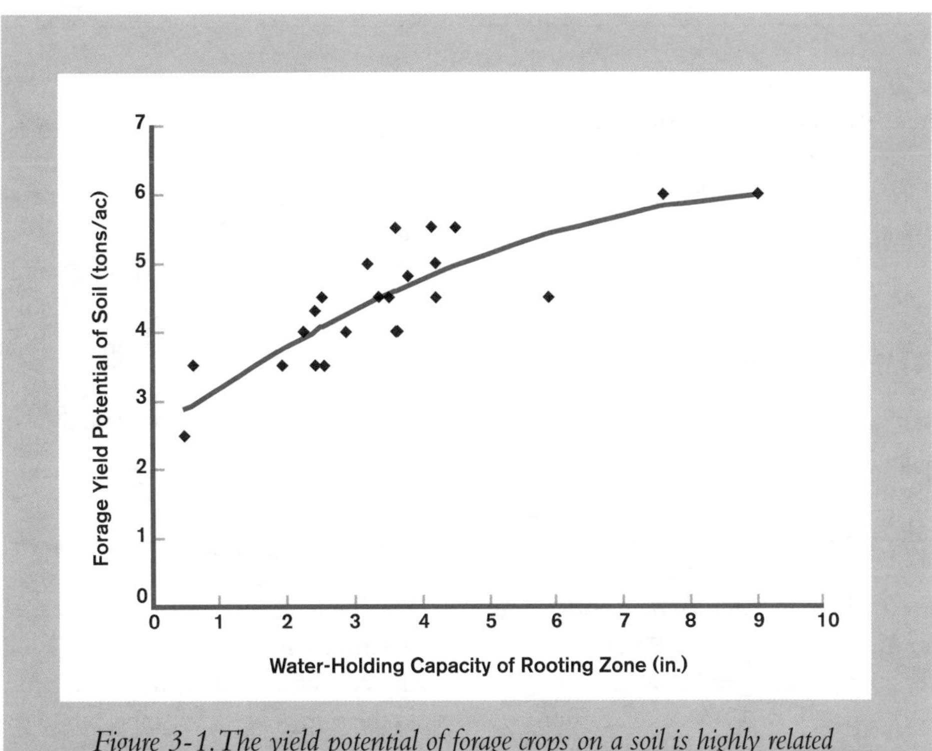

Figure 3-1. The yield potential of forage crops on a soil is highly related to the soil's water-holding capacity, which is a function of the rooting depth of the soil and the water-holding capacity per unit of soil depth.

Table 3-1. Realistic yield expectation for clover-grass and alfalfa-grass hay (tons/ac/yr) and grass-clover pasture (animal unit days of grazing or AUD/ac/yr) under good management, as determined by soil type for selected West Virginia soils.

Soil	Slope (%)	Clover-grass	Alfalfa-grass	Grass-clover pasture (AUD/ac/yr)
		–(tons/ac/yr)–		
Albright silt loam	8–15	3.0	3.5	200
Berks shaly silt loam	3–8	3.0	3.5	200
Berks shaly silt loam	15–25	2.0	3.0	170
Clarksburg channery silt loam	3–8	3.0	3.5	200
Clarksburg channery silt loam	8–15	3.0	3.5	200
Dekalb channery loam	5–12	2.8	3.2	120
Dekalb channery loam	12–25	2.6	3.0	105
Ernest silt loam	8–15	3.0	3.5	140
Frankstown silt loam	3–10	3.5	4.5	165
Frankstown silt loam	10–20	3.2	4.3	160

these published levels by 25% or more. Where improved management is used, crop yield records for the farm are more accurate than general areawide estimates.

SOIL FERTILITY

Soil fertility is a measure of the nutrient status of the soil. Soil testing enables the manager to evaluate the soil's fertility. Soil testing labs regularly provide indices of plant available phosphorus (P), potassium (K), calcium (Ca), magnesium (Mg), and acidity (soil pH). These major minerals may be reported as the amount of mineral element (i.e., P, K, Ca, or Mg), or the fertilizer oxides or carbonates; phosphate (P_2O_5), potash (K_2O), calcium carbonate ($CaCO_3$), and magnesium carbonate ($MgCO_3$). By special request, other analyses can be conducted for soil organic matter and trace elements.

Soil testing is only a chemical analysis. The value of the index for a nutrient depends upon the laboratory method used to extract the plant nutrient from the soil. Laboratories using different extraction methods can have very different index values for the nutrient content of a soil. When comparing values from different labs, the relative readings of low, medium, high (or optimum), or excessive should be approximately the same.

PHILOSOPHIES OF SOIL TESTING

Three philosophies are commonly used to determine a soil fertility program. They are the maintenance approach, the cation saturation ratio, and the sufficiency level approach to soil fertility (6).

The maintenance approach is to apply a quantity of nutrients adequate to replace the nutrients expected to be removed by the crop regardless of the soil test level. This approach is appealing because a soil's nutrient supply capacity is conserved, but this doesn't consider the capacity of soils to supply nutrients. There are situations in which the soil's delivery capacity may be adequate for top yields for a period of years without the application of any additional nutrients. This philosophy might be described as an insurance approach by overfertilization. However, using the maintenance approach without regard to the nutrient-supplying power of the soil is economically wasteful and potentially environmentally harmful.

The cation saturation ratio concept proposes that an ideal soil has a distribution of exchangeable cations that is most beneficial to plant growth. This proposed ratio is 65% Ca, 10% Mg, 5% K, and 20% hydrogen (H). Many studies have shown that this system is not valid across soil types. This concept does not consider nitrogen (N), P, sulfur (S), or micronutrients. Research has shown that soils with wide variations in those ratios have had no adverse effects on yields or crop quality (6).

The sufficiency level approach is based on long-term calibration of soil tests with field yield response data. Studies have shown that above a certain soil test level, no yield response will occur to an applied nutrient. The soil fertility ranges of "response assured," "response likely," "response possible," and "response unlikely" are commonly shown on soil test results as very low, low, medium, and high nutrient level, respectively. This concept is supported by most of the U.S. university soil testing laboratories (6).

HOW TO TAKE A SOIL SAMPLE

Soil testing is meaningful only if the soil sample accurately represents the field from which it is taken. If the sample is contaminated with lime, manure, or fertilizer while being collected or prepared, the results will be incorrect. If a soil sample does not represent the area, then the fertilization recommendations will be meaningless.

Secure a soil test packet from the local extension office or commercial laboratory. The packet will contain an information sheet, a plastic bag in which to place the soil sample, and a mailing container.

When taking a soil sample, first select the area to be sampled. This is usually a small field or part of a field. If the field has areas that differ greatly in soil texture, slope, topographic position, or drainage, take separate soil samples from each of these identified areas.

If there are areas within a field that have had different management over the years, such as cropping patterns, manure applications, or animal wintering areas, take separate soil samples from each of these areas.

Avoid sampling from unusual areas such as wet spots, eroded banks, bare spots, back furrows, field edges, near trees, hedgerows, or livestock watering or mineral feeders. Do not combine soil cores from areas that were managed differently.

Use a soil sampling tube or auger when taking a soil sample. These tools are punched or twisted into the soil to the proper depth to extract a small soil core. These tools may seem expensive, but they are inexpensive compared to the cost of fertilizer and lime and make it easier to obtain a better sample.

Take a minimum of 20–30 randomly selected soil cores with the sampling device (assuming a field size of 10 acres or less). Divide large fields into smaller areas based on prior management, soil type, and slope position.

How Deep to Sample

How deep to take a soil sample depends on field management. Fields in continuous hay or pasture will have had fertilizer and lime surface-applied, with plant residues and manure returning nutrients to the soil surface. These soils will have the highest nutrient content in the upper few inches and should be sampled to a 2-inch depth. This will provide the most sensitive measure of soil fertility change with fertilizer and lime applications. For permanent hay fields and pastures, take the soil cores by brushing aside any coarse organic matter from the surface and taking the cores to a 2-inch depth. If you want to evaluate the fertility status of deeper rooting zone soil, take a second soil sample using cores from the 2–8-inch depth.

Depending on the tillage system used, fields in a crop rotation may have lime and fertilizer incorporated into the soil. In no-till systems, there is little incorporation. If the land is chisel plowed or disked, there will be some incorporation in the top 2–4 inches of soil. If the field is occasionally moldboard plowed, there will be deep incorporation of lime and fertilizers. For meadows in crop rotation, brush aside coarse organic debris from the soil surface and take a soil core to the effective incorporation depth of the tillage method used—usually 4 inches for chisel plowing and disking and 6–8 inches for moldboard plowing. On fields managed under no-till, take two samples, one to 1-inch deep and the second from 1 to 6 inches deep.

When to Sample

Soil samples can be taken any time of the year when the soil is moist enough for the tube or auger to be used but dry enough so that the soil will not compress and be difficult to dry and mix. Soil samples should not be taken when the soil is wet or frozen or shortly after a drought or lime or fertilizer application.

Sampling soils that were recently limed usually produces misleading soil test results. Most laboratory methods used in the Northeast are based on an acidic extraction of nutrients from the soil. A limed soil can have free calcium carbonate in it for up to 3 years after liming. This free calcium carbonate neutralizes the mild acid extraction, resulting in a lower P extraction than would happen if no free carbonate were present. This low P extraction results in a higher recommendation for added phosphate fertilizer than needed.

Soil fertility indices do change over the year due to weather and plant activity. Therefore, it is best that soil samples be taken the same time of year, fall or spring, when evaluating the effectiveness of a soil fertility program.

Handling the Soil Sample

Collect the soil cores from a sampled area in a clean plastic container. Do not combine soil cores from areas that were managed differently. If the cores are wet, spread them out on clean paper and let them air dry in a protected, shaded place. Never oven-dry soils, as this will adversely affect the test results. When dry, gently crush the cores and mix well. Remove rocks and coarse organic matter such as roots from the sample. Take a subsample of the mixed soil (about 1 pint) for submission to the lab for analysis. Use the bag and container supplied with the soil test kit.

Soil Test Information Sheets

Fill out the information sheet as completely as possible for each sample. Additional information or questions may be included in the remarks section. Send the soil samples and soil test information sheets to the selected lab. Soil samples should be collected and mailed well in advance of management planning and implementation.

How Often to Soil Test

The frequency of soil testing depends on the crop grown, previous fertilization rates, and timing of lime application. In general, every 3–5 years will suffice for forage crops. Some recommend that for continuous row crops and alfalfa or if high fertilization rates are being used, soil fertility should be tested every 1–2 years.

THE SOIL TESTING LABORATORY REPORT

The soil test report is used in making fertilization and liming decisions. Different labs may list the report information in a different order than discussed here.

The first part of the report contains information that identifies the soil sample with the sample location. This information includes the field name, size, and previous crop grown as submitted on the information sheet sent in with the sample.

The second part of the report contains the soil test results. This includes soil pH, P, K, Ca, and Mg content and the sample's cation exchange capacity, base saturation, and lime requirement. If requested, the organic matter content and/or the micronutrients zinc, copper, and manganese may be listed in this section.

Depending on the lab, the soil test result values are given in pounds per acre or parts per million. Although these nutrients are often referred to as available nutrients, the numbers in the report are only indexes of the nutrient content of the soil. Depending on the chemicals used to extract the soil sample ("extracting" pulls elements off the soil particles for measurement), the values for soil test P may differ by a factor of 10 between two labs. The number from one lab is as good as the other. It is the correlation of the index to crop yield in the field that determines the accuracy of recommendations.

In addition to the numerical index values for the nutrients, soil test values are classified as being "Low," "Medium," "High" or "Optimum," and "Very high" or "Excessive." The implications of each of these categories are explained below.

Low—There is a high probability that the plant nutrient will limit plant growth and yield. The fertilization recommendation made will increase the level of that nutrient in the soil. If the level is very low, several years of fertilizing or liming may be necessary to reach "High" or "Optimum." For row crops, banding fertilizer near the row may be critical to ensure efficient use and nutrient availability to the crop.

Medium—There is a good probability that the nutrient may limit plant growth or yield by the end of the growing season in years of very good growing conditions. Fertilizing or liming is recommended in moderate amounts to cause a slight increase in soil test level after the crop has been harvested or to support exceptional yields in a very good year.

High or Optimum—The plant nutrient is in the ideal range to support optimum plant growth and yield. Any fertilizers applied are to compensate for crop removal so that soil fertility may be maintained. For row crops a small amount of starter fertilizer containing the nutrient may be recommended to give the crop a good start.

Very high or Excessive—The plant nutrient level is higher than needed to support optimum crop growth and production. Growth and yield may be inhibited by the soil nutrient that is testing in this range, because excess of one nutrient may interfere with the uptake or availability of other nutrients. With surface-applied nutrients on permanent sod, "Very high" may be an environmental risk due to losses to the environment.

Soil pH is an indicator of the acidity of a soil. A pH below 7.0 is acidic and pH above 7.0 is alkaline. Soil pH affects the availability of most plant nutrients. Different crops require different soil pH levels for optimum production. Many row crops, small grains, and grass-clover mixtures grow best at a pH of 6.0 or above. Alfalfa requires a slightly higher soil pH of 6.5 or above. Pure grass stands do well with a pH of 5.6 and above when adequate fertilizer N is applied.

Crop production may be severely reduced in soils with a pH at or below 5.0. As soil pH falls below 5.5, available aluminum and manganese increase. As pH drops below 5.0, they may begin to approach toxic levels. Also, P and molybdenum availability decrease as soil becomes more acidic. As aluminum increases in soil, it binds with P so that the P is not available to plants.

Lime requirement (LR) is determined in a buffered solution. The LR is the amount of pure calcium carbonate (100%-effective neutralizing value) needed to raise the soil to pH 6.6 when mixed well into the top 6 inches of the soil. Some labs will adjust recommendations for LR for different crops and different depths of lime incorporation.

The last part of the soil test report shows the recommendations for use of agricultural lime and fertilizers. These recommendations are based on soil test values, yield goals, and estimated nutrient removal by the crop. Recommendations for N fertilizer are not based on soil testing. N recommendations are based on expected crop and soil RYE and fertilizer use efficiency. If the RYE is different from that assumed in the soil test report, the recommended fertilizer rates for all nutrients need to be corrected for the local conditions. Credit should be given for any manure or other organic matter added to the soil and contributions from legumes such as alfalfa or clover in the sward.

Fertilization recommendations for P and K are given in pounds per acre of the fertilizer oxides phosphate (P_2O_5) and potash (K_2O). The pounds of fertilizer needed per acre must be calculated based on the nutrient content of the fertilizer used (see table 3-2). Commercial fertilizer packages

Table 3-2. Commercial fertilizers with their fertilizer nutrient content per 100 pounds of fertilizer, soil acidifying effect, and salt index.

	N	P_2O_5	K_2O	Acidifying effect	Salt index
Ammonium nitrate	33	0	0	1.8	104.7
Ammonium sulfate	21	0	0	5.35	69.0
Anhydrous ammonia	82	0	0	1.8	47.1
Calcium nitrate (produces basic soil reaction, not acidic)	15.5	0	0	(1.35)	52.5
Urea	46	0	0	1.8	75.4
Urea ammonia nitrate solutions	28–32	0	0	1.8	N/A
Triple superphosphate	0	46	0	0.7	29.9
Diammonia phosphate	18	46	0	N/A	7.8
Superphosphate	0	20	0	0	10.1
Potassium chloride	0	0	60	0	116.3
Potassium sulfate	0	0	53	0	46.1
Sulfate of potashmagnesia	0	0	26	N/A	43.2

state the ratio of pounds of N, phosphate, and potash (N-P_2O_5-K_2O) contained per 100 pounds of fertilizer. Let's say we're using a 0-45-0 fertilizer. The fertilizer tag or package tells us there are 45 pounds of phosphate per 100 pounds of fertilizer or 0.45 pounds of phosphate per pound of fertilizer. If we need 60 pounds of phosphate per acre, we divide 60 by 0.45 to find that we need 133 pounds of 0-45-0 fertilizer to provide the 60 pounds of phosphate (60/0.45 = 133).

PLANT RESPONSE TO SOIL FERTILITY AND pH

To determine the economics of a crop's fertilizer requirement, the manager needs to know the crop's response to soil fertility and to added fertilizer. Forage crop response to soil test P and K is shown in figures 3-2 and 3-3 (p. 40) (4, 9, 10).

These response curves are based on research in Ohio and New York with the soil test values calibrated to the West Virginia University (WVU) soil test system (Mehlich 1). These response curves show forage yields over a range of soil test results when no fertilizer is applied. Relative yields are expressed as a fraction of the maximum yield obtained with high rates of fertilization. Expression of crop growth as relative yield helps account for different crop and soil RYE and yearly weather variations. When soil test P is greater than 50 (figure 3-2) or soil test K is greater than 125 (figure 3-3, p. 40), crop yield is not increased by fertilization with phosphate or potash. Soil test levels in this range correspond to the lower end of the "High" or "Optimum" categories. For different labs these index numbers will change, but the general estimate of "Low," "Medium," or "High" will be about the same.

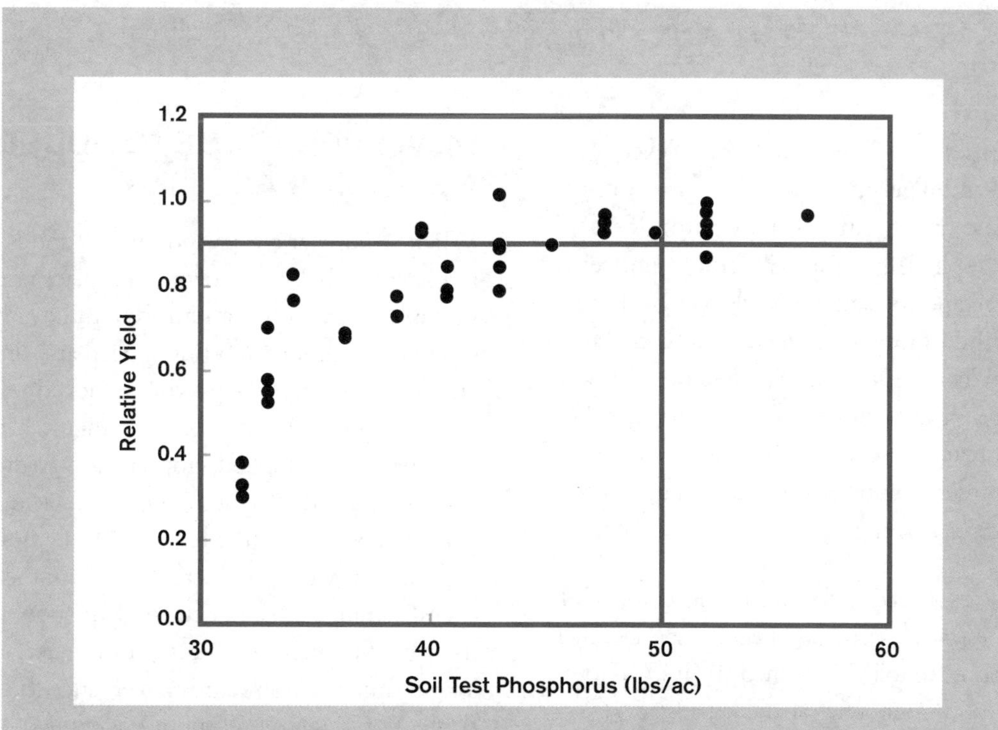

Figure 3-2. Relative yield is low at a low P soil test index but nears the crop's maximum yield potential when the P soil test index is at its optimum.

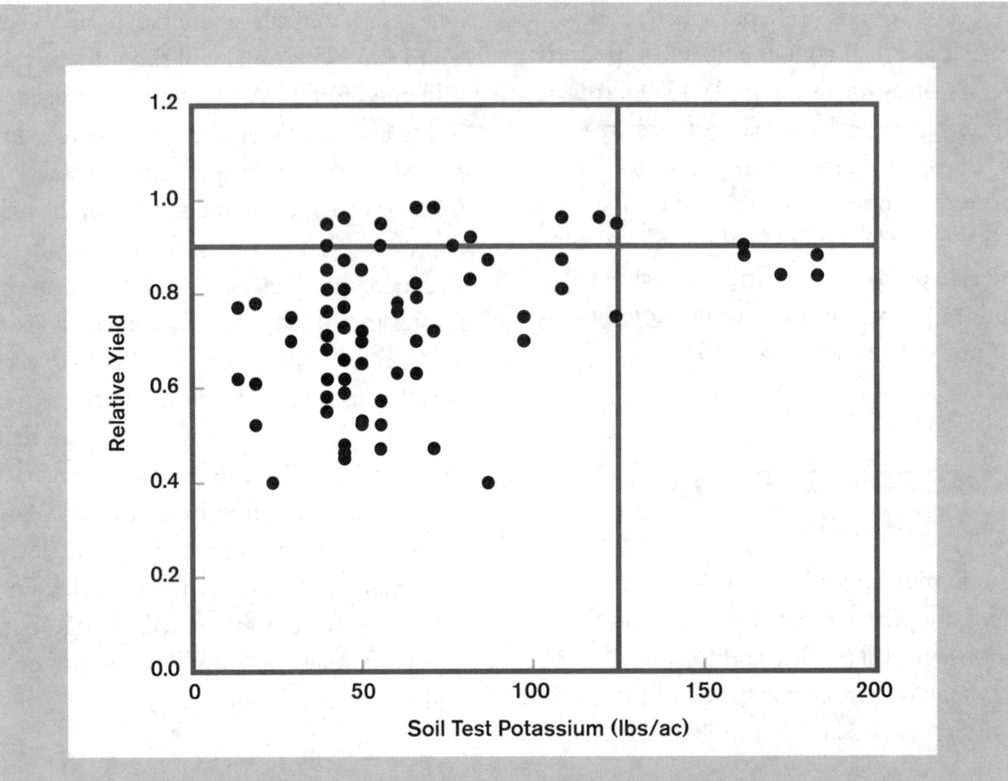

Figure 3-3. Relative yield is low at a low K soil test index but nears the crop's maximum yield potential when the K soil test index is at its optimum.

Plant responses to different soil fertility levels are for short-term experiments and do not account for removal of plant nutrients in the crop. If no fertilizer is applied, the crop will remove nutrients from the soil and the soil test levels will go down. Over a period of time crop yields will then also go down. The decrease in yield will depend on the amount of nutrients removed in the harvested crop, the nutrient reserves in the soil, the release rate of soil reserve nutrients, and the length of time that fertilizer is not applied.

Forage legumes should be the main contributor of N in an economical forage system. Legume species differ in their response to soil pH (figure 3-4) (10). On low-pH sites, clover, birdsfoot trefoil, and lespedezas maintain production better than alfalfa does.

PLANT RESPONSE TO ADDED FERTILIZER AND LIME

When looking at crop response to P and K fertilizer, it is helpful to express fertilizer rates as a multiple of the nutrient removed in the crop. If the soil test is optimum and P is applied at 1 times the crop removal rate (P applied/P removed), crop yields are 90–100% of maximum (figure 3-5) (10). If no fertilizer is applied, crop yields may drop to between 50 and 80% of the potential yield over a period of 3–7 years (the length of time these experiments were conducted). If P is in the excessive range, as can occur on large poultry and dairy confinement operations, it may take a much longer time for crop removal to reduce the nutrient content of these soils. If more P is applied than the removal rate, there will be no yield increase. Soil test levels will increase but will return no economic

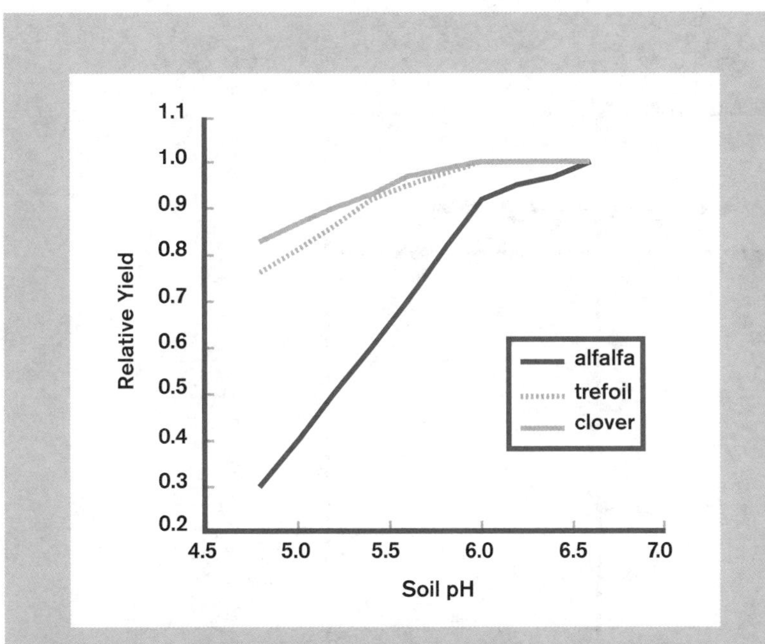

Figure 3-4. Legumes differ in their yield response to soil pH. Alfalfa requires a pH above 6.5 for maximum production, but clovers and birdsfoot trefoil need a pH only above 6.0.

value unless fertilizer inputs are reduced in the future to draw on this reserve fertility.

For K the effect is similar (figure 3-6, p. 42) (10). When the soil test is high or optimum, an application rate of 1 times the removal rate (K applied/K removed) maintains crop productivity at near the maximum level. Again, with no added K, the crop yield drops to 30–90% of potential over a 3–7-year period. The drop in yield without K fertilizer is determined by soil type, crop yield, and forage species. Some soils contain clays that hold high levels of reserve K. In these soils, soil test values for K will stay higher longer.

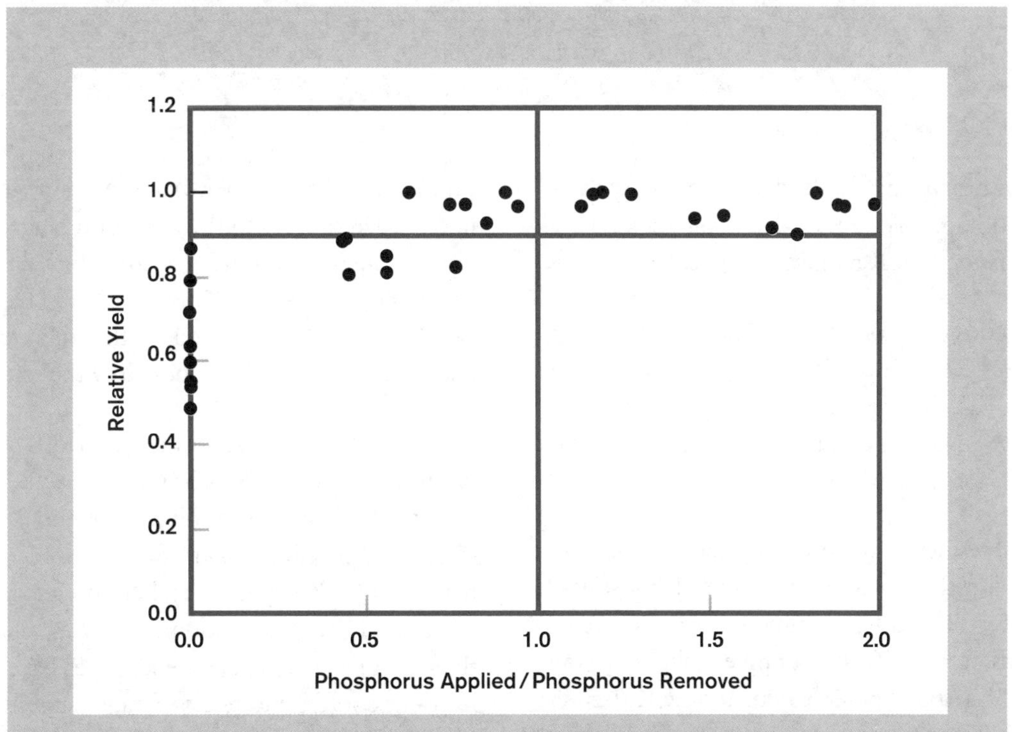

Figure 3-5. At a medium to high P soil test index, applying phosphate at the crop removal rate will keep forage production near the maximum and maintain the soil test index.

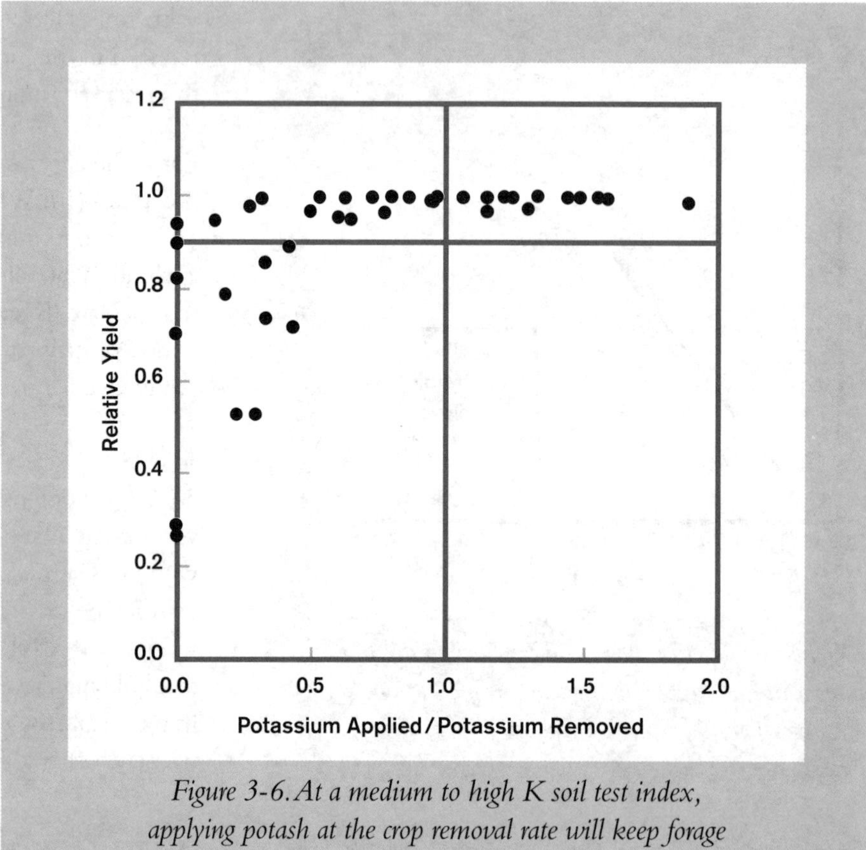

Figure 3-6. At a medium to high K soil test index, applying potash at the crop removal rate will keep forage production near the maximum and maintain the soil test index.

On sandy soils that have little clay or on soils with clays that do not hold reserve K, the soil test level will go down faster with K removal by the crop.

Another consideration in K fertility is that legumes are less competitive than grasses in the uptake of K. When managing for legumes, it is essential to maintain optimum K levels in the soil to provide for their needs (1).

When soil test levels are low, crop yields will be greatly reduced if fertilizers are not used. For soils testing low in P and K, when no fertilizer is applied crop yields drop to 10–40% of potential due to low P and to 30–70% of potential due to low K (figures 3-7 and 3-8, p. 43) (10), respectively. For soils testing low in P, the cost to get near potential yield increases because the fertilizer rate needed is 2–4 times removal rate rather than the 1 times removal rate needed on soils in the optimum range. On low-testing soils, fertilizing at the removal rate gives a yield between 40 and 100% for P and between 80 and 100% for K, depending on the soil's ability to release these minerals and how low the test value is.

The rate of nutrient removal by various crops is provided in table 4-3 (p. 53). Pasture and hay crops remove 9–11 pounds of phosphate and 38–51 pounds of potash per ton of hay equivalent harvested (90% dry matter). Legume crops such as alfalfa are heavier users of nutrients, especially K, than are grass crops because legumes have a higher nutrient content per ton of forage harvested. High-yielding crops such as alfalfa and corn silage are heavier users of nutrients because they have high yields per acre.

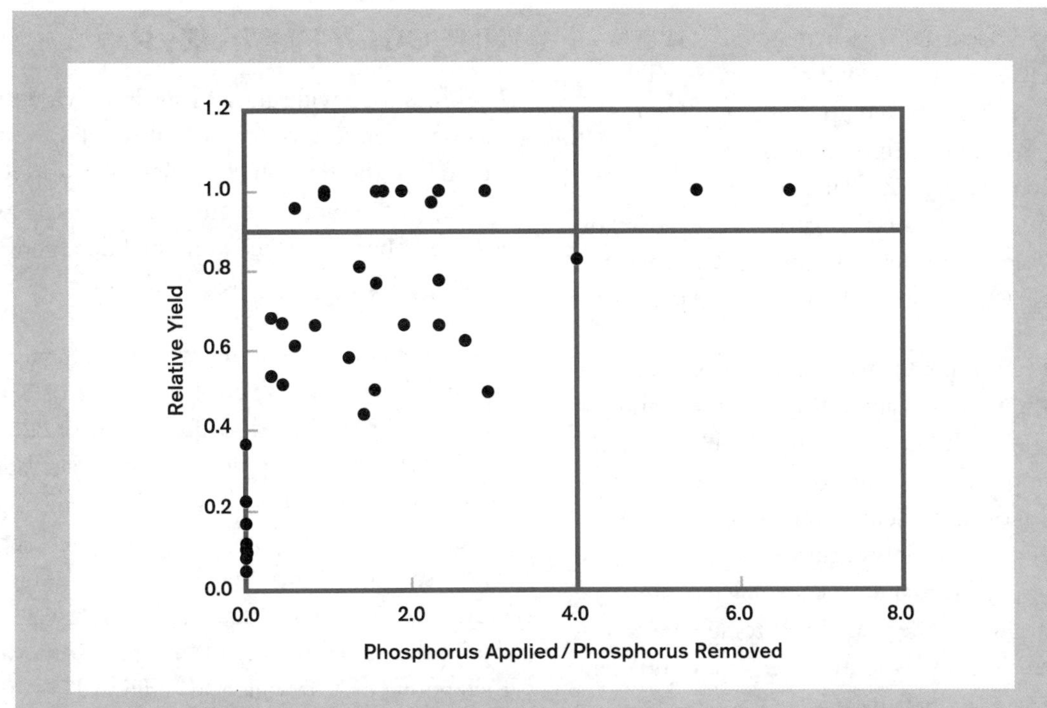

Figure 3-7. At a low P soil test index, phosphate must be applied at two to four times the crop removal rate to allow yield to be near maximum and to increase the P soil test index.

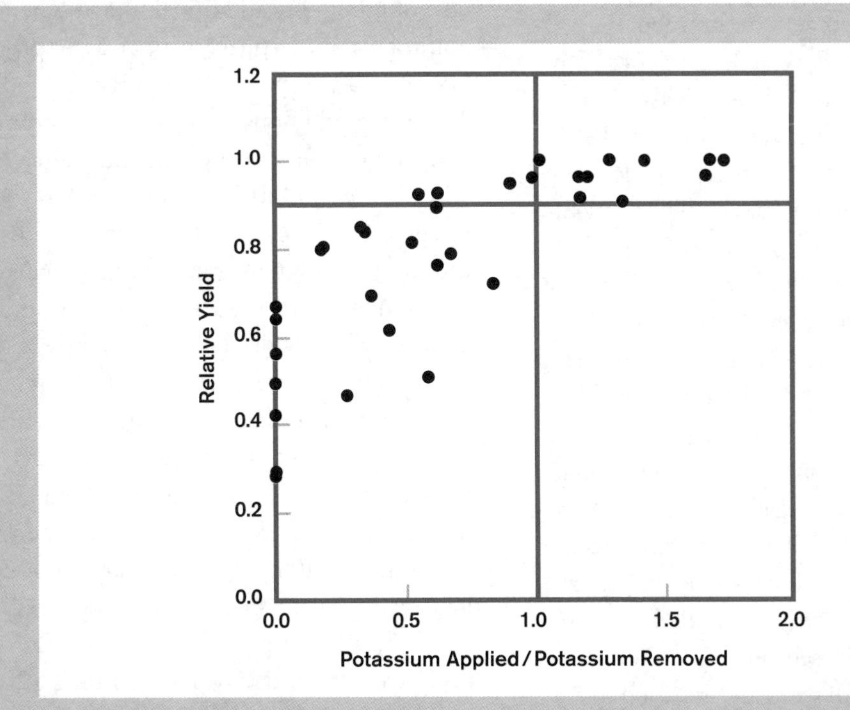

Figure 3-8. At a low K soil test index, potash applied at the crop removal rate will keep forage production near the maximum and maintain a low K soil test index. Additional K fertilization would be needed to increase the K soil test index.

If the crop is moved off the farm (sold or fed somewhere else), purchased fertilizer is needed to replace the nutrients taken up by the crop. On the other hand, if feeds are moved onto the farm (purchased or brought in from another farm) and the resulting manure is properly managed and applied on the farm, these imported nutrients will increase soil fertility and reduce the need for purchased fertilizers.

Nutrient removal rates are highest for pastures, but most of the nutrients are returned in the manure and urine deposited back on the soil. However, if the return of manure is not uniform because the cattle are allowed to deposit much of the manure under trees, a major loss of nutrients can occur from the open areas and excess nutrients can accumulate under the trees. Little forage tends to grow there, so there is an increased risk of nutrient loss to surface and groundwater.

Plant nutrients can be provided to a field from manure, compost, or commercial fertilizers. Table 3-2 (p. 38) lists the nutrient content of common commercial fertilizers along with the fertilizer's acidifying effect and salt index. This table does not include specialty organic fertilizers with limited availability. Specific cost and analytical values for these products can be obtained from dealers. The nutrient content is pounds of nutrient per 100 pounds of fertilizer, as shown on the fertilizer tag. The fourth column lists the soil acidifying effect of the fertilizer in pounds of calcium carbonate required to neutralize the acidity per pound of N applied. Ammonium sulfate has the highest acidifying effect but supplies sulfur, which may be needed in some areas. Calcium nitrate has a liming effect on soils (negative acidifying effect) because it is a carrier of Ca. The fifth column lists the salt index of these fertilizers. High salt index fertilizers should not be used at planting near seeds in the soil because they can burn the seedlings' roots. They should not be mixed with seeds for broadcast seeding. Triple superphosphate and superphosphate are the only fertilizers that are safe for mixing with seeds for broadcast seeding.

NITROGEN FERTILIZER

Legumes used with grasses usually provide the most economical source of N through biological N fixation (figure 3-9) (10). They also improve the quality of the forage, because legumes are lower in cell wall fiber (measured as neutral detergent fiber), which limits intake of high-forage diets.

When commercial N fertilizer is used on grass hay or pasture, the response depends on the rate of N used, soil quality, weather, and adequate levels of other nutrients (figure 3-10, p. 46) (3). When calculating the economics of N application, one should use the forage production response per pound of N applied. This response is shown in figure 3-11 (p. 46) for the same data set shown in figure 3-10. In this example we see that at 100 pounds of N per acre or lower, each pound of N provided an additional 20–40 pounds of forage dry matter.

We can use this response to estimate the marginal cost of producing additional forage when using commercial N fertilizer. If N costs $0.30 per pound and we apply 60 pounds of N and obtain 30 pounds additional dry matter per pound N, then the marginal cost of the forage dry matter is $0.01 per pound (0.30 $/lb N ÷ 30 lb additional forage) or $20 per ton. If the site is a grass meadow with little legume content and we want to extend the grazing season, each pound of N applied will provide about 20 pounds of dry matter. This represents about 1 day of grazing for a 1,000-pound animal eating 20 pounds of dry matter per day. For example, if a manager applied 1 ton of urea (46% N) to a 20-acre field (100 lb urea/ac or 46 lbs N/ac) there would be an additional 18,400 pounds of forage produced (46 lbs N/ac x 20 acres x 20 lbs dry matter/day). A herd of 30 1,200-pound cows eating 2% of their body weight per day would consume 720 pounds/day (30 cows x 1200 lbs/cow x 0.02 body wt = 720 lbs/day). This rate of N on the field, if grazed properly, would provide 25.6 days of additional grazing for the herd (18,400 lbs forage/

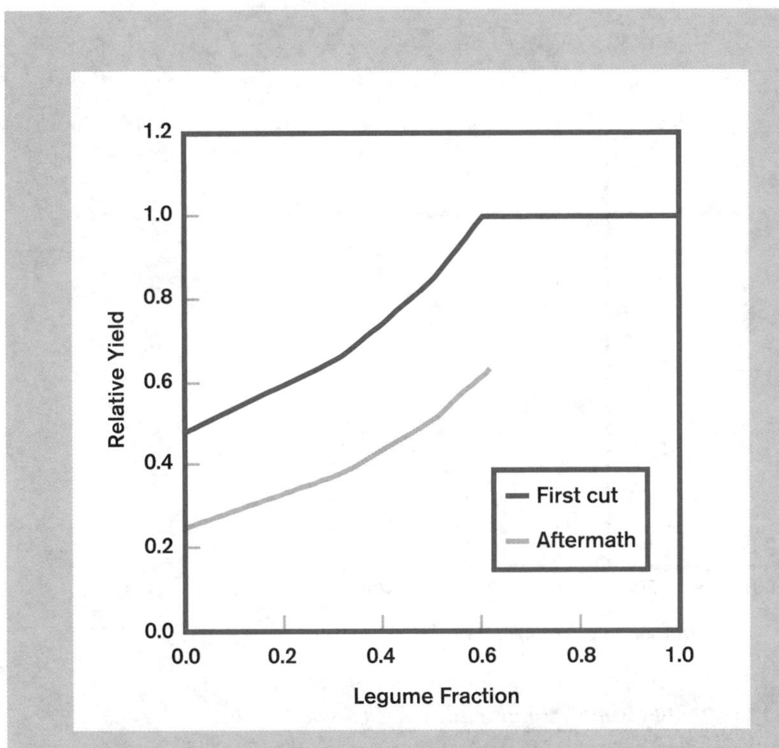

Figure 3-9. Forage growth of a grass-legume mix depends on N fixed by the legume. In an older stand where soil organic matter is well developed, maximum forage yield may be achieved at around 50% legume content. However, for a new seeding on previously cropped soils with low soil organic matter, maximum yield may require a much higher legume content in the stand.

720 lbs forage/day = 25.6 days). When there is a reasonable percentage of legume in the field (20–30%), the response to N fertilization will be about half of what is shown in figure 3-11 (p. 46) (5).

The general recommendation for N fertilization of grasses for hay is to apply 150–180 pounds N per acre per year as plant-available N from poultry litter, livestock manure, or commercial fertilizer. When using commercial fertilizer, N should be split-applied at 50–60 pounds per acre, three to four times during the growing season. Lower N rates are recommended for pastures because there is considerable recycling of N in pastures due to the return of manure and urine by livestock and an increase in soil organic matter that serves as a pool of N for plant growth. The best source of N in pastures is legume N fixation. Where legumes are lacking, applying N at 50–100 lbs per acre per year may improve forage production.

There are risks in the use of fertilizer N due to weather. When using urea, a lack of rain within a few days of application will result in much of the urea being converted to ammonia and volatizing off the field. Using ammonium nitrate or ammonium sulfate, which do not break down and volatilize, eliminates this problem. Another risk on poorly drained soils is the loss of N through denitrification (14) during wet spring weather.

From an economic point of view, it is difficult to justify commercial N application as a general management practice on pastures and meadows. Managing soil fertility and pH and sward defoliation to improve legume production is often a better use of financial resources. However, strategic N applications can pay off, especially when applied in late summer to stimulate growth for fall and winter grazing.

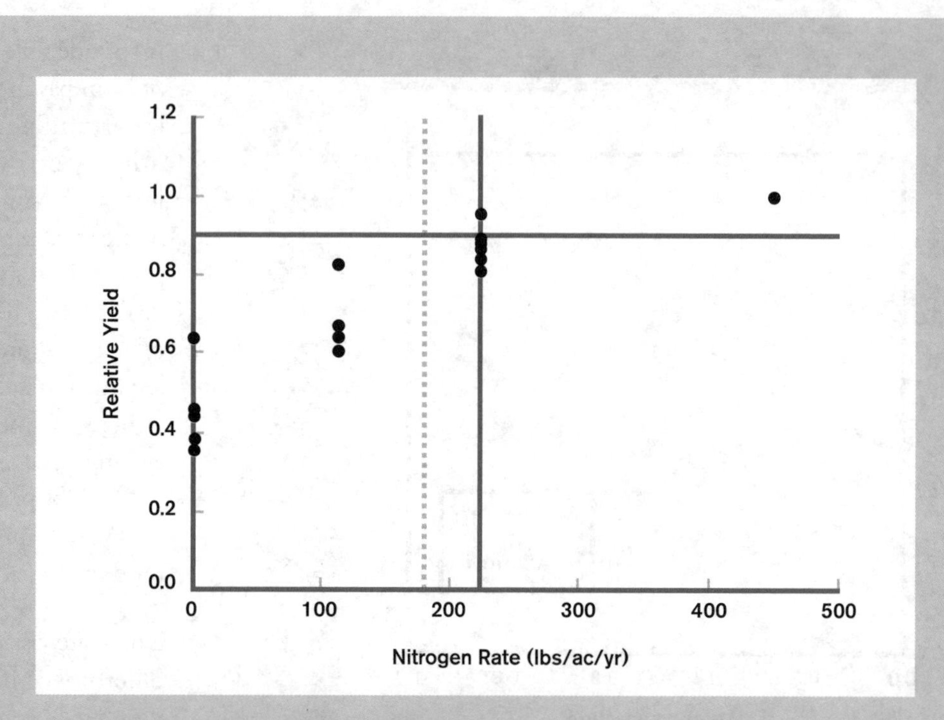

Figure 3-10. Under West Virginia climatic conditions forage grasses such as orchardgrass, tall fescue, and timothy reach about 90% of their potential hay yield when N is applied at about 220 lb/ac/yr. There is increased risk of nitrates leaching to groundwater as N rates go above 180 lb/ac/yr (dashed line).

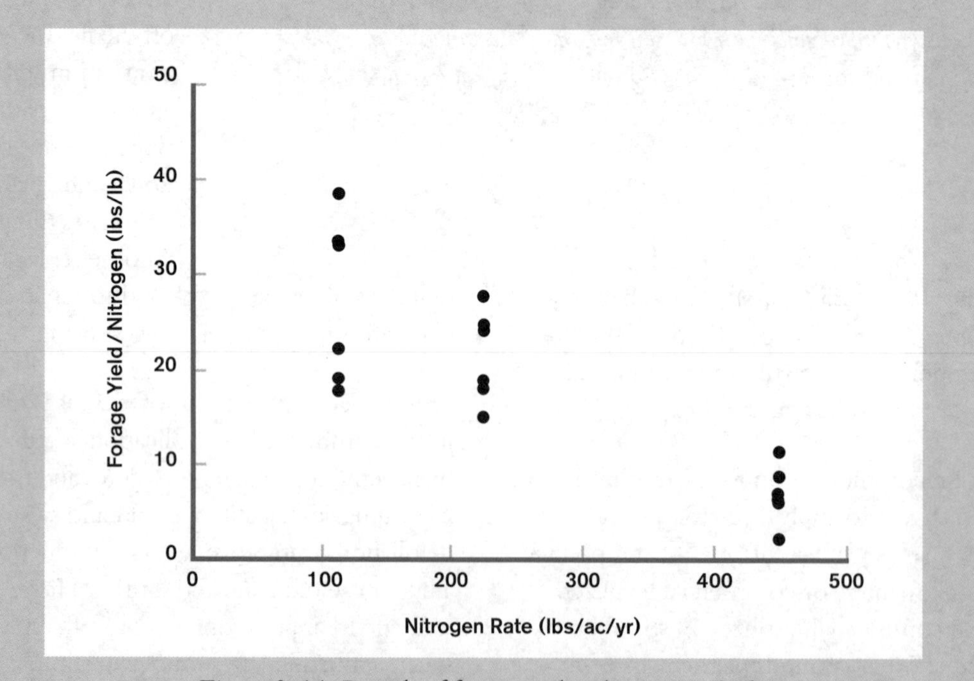

Figure 3-11. Pounds of forage produced per pound of N applied decrease as the rate of N increases.

LIMING RATES

The lime requirement reported in the soil test report is for raising the soil pH to that needed by alfalfa (6.6). For grass and clover a soil pH of 6.0 or above is adequate. The indicated liming rate is usually for lime incorporated into the soil.

When lime is surface-applied, the soil pH will peak in 3–4 years, but the pH will be maintained in the top 2 inches for a relatively long time (figure 3-12) (2). When agricultural lime is applied and incorporated into plowed soil, the pH increases rapidly and peaks within 12 months within the 6-inch plow layer (figure 3-13, p. 48) (8). At high liming rates the peak pH will hold for a while and then begin to decrease. At low liming rates the pH will begin to drop off soon after the first year.

When liming for acid-tolerant legumes during establishment, apply one-half the recommended lime after primary tillage and lightly disk it into the surface. For no-till establishment or pH maintenance using surface applications for acid-tolerant legumes, limit lime to 2 tons per acre and soil test again after 3 years to check for additional lime needs.

How much lime to apply per acre for a given level of pH adjustment is determined by the selected limestone's calcium carbonate equivalent and fineness of grind. State extension services have fact sheets that show how to compare different

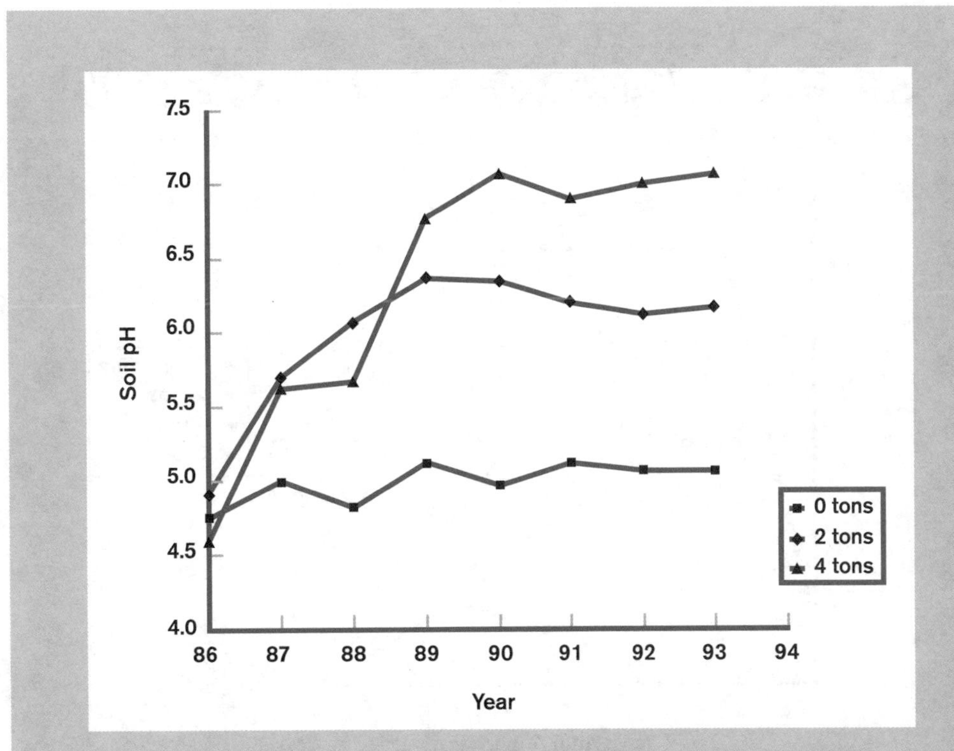

Figure 3-12. When lime is surface-applied the change in soil pH is relatively slow in the top 2 inches of the soil, but the change is relatively long lasting. Two tons of lime were applied in 1986 to the 2- and 4-ton treatments. An additional 2 tons were applied to the 4-ton treatment in 1988.

Source: Adapted from Bryan, W.B., and K.C. Elliot. 1990. Effects of sample depth, and of lime and phosphorus applications on soil test levels in pasture soils. In R.J. Write, V.C. Baligar, and R.P. Murrmann (ed.). Plant-Soil Interactions at Low pH. *Kluwer Academic Publishers, Boston.*

agricultural liming materials for their cost-effectiveness based on these factors. One example using the label required by law in West Virginia can be found on the Internet at HTTP://WWW.CAF.WVU.EDU/~FORAGE/3212.HTM. For states with different lime labeling laws, check with your local extension service for similar calculations using the values from the label required by your state.

Table 3-3 provides a listing of different liming sources and their calcium carbonate equivalent (CCE) content. Calcium sulfate (22% Ca) is not a liming source but does provide Ca for crops where an increase in soil pH is not desired. When purchasing agricultural lime be sure to get a copy of the product label so you'll know its CCE and fineness of grind.

Mg is recommended if the soil tests "Low" or below 10% of the cation exchange capacity of a soil. The most economical source of Mg is dolomitic (high Mg) limestone. Quite often when lime is not needed, no Mg recommendation is made. Table 3-4 lists common sources for Mg and their CCE content, when applicable. Dolomitic limestone has a CCE greater than 100% because a Mg ion is lighter than a Ca ion so there are more molecules of magnesium carbonate than calcium carbonate per pound of lime.

Cases of acute Mg deficiency in crops are rare. In most cases, it is acceptable to wait until lime is needed and then apply dolomitic lime. Where forage is grown on soils testing low in Mg, cattle should be fed a high-Mg mineral mixture to

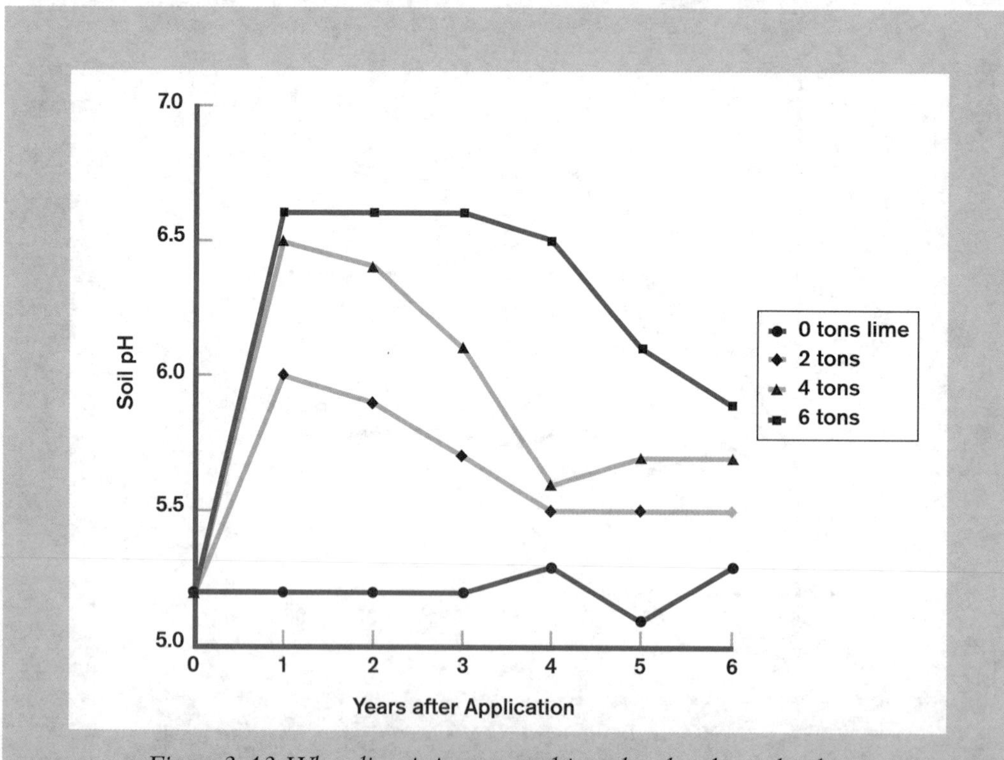

Figure 3-13. When lime is incorporated into the plow layer the change in soil pH is relatively fast in the 6–8-inch plow layer depth, but the pH change is relatively short-lived for low rates of lime.

Source: Adapted from Reid, W.S., W.L. Griffeth, R. Feuer, and R.B. Musgrave. 1967. Effect of Lime Rate, Timing, and Placement on the Yields of Corn, Oats, and Hay. *Agronomy Mimeo. 67-13. Agronomy Department, Cornell University, Ithaca NY. 10 pp.*

Table 3-3. Liming sources and their calcium carbonate equivalent (CCE).

Source	CCE
Burned lime	150
Hydrated lime	120
Dolomitic limestone	104
Ground limestone	75–95
Marl/oyster shell	95
Calcium sulfate	22% Ca

Table 3-4. Mg sources and their content of magnesium and CCE, if applicable.

Source	CCE
Dolomitic limestone	105 CCE
$MgCO_3$ (magnesium carbonate content)	42% Mg
MgO (magnesium oxide content)	20% Mg
Mg (magnesium content)	12% Mg
$CaCO_3$ (calcium carbonate content)	54% (% of Mg as $CaCO_3$ in dolomitic limestone)
Magnesium oxide	
MgO	55–60% Mg
Magnesium sulfate	
$MgSO_4$	9–22% Mg

prevent grass tetany. Research shows that soil P is as important as Mg in ensuring adequate uptake of Mg by plants for use by ruminants (12, 13). To minimize the risk of grass tetany, do not apply high rates of N or potash to pastures in early spring. Fertilizers and manures containing high K or N content reduce the availability of Mg to animals.

CROP RESPONSE TO ADDED NUTRIENTS

Crop response to soil test values and added fertilizer has been discussed in terms of relative yield. We can estimate the expected crop yield for a soil by multiplying the relative yield of the most limiting nutrient by the soil RYE for the crop of interest. For example, an Ernest series soil has a RYE of 3.0 tons per acre for clover-grass hay (table 3-1, p. 34). If the current soil test report gives a low P value and the relative yield due to soil test P is about 0.50 (from figure 3-2, p. 39), we can estimate current expected yield without fertilizer P as:

$$\text{Expected yield} = \text{Relative yield} \times RYE_{\text{clover-grass hay}} = 0.50 \times 3.0 = 1.5$$

Remember that we discussed above that pasture and hay crops remove 9–11 pounds of phosphate per ton of hay equivalent harvested. If we choose to apply phosphate at four times the removal rate to ensure top yield and an increasing soil test value (figure 3-5, p. 41), we need to apply:

$$9 \text{ lb } P_2O_5/\text{ton hay} \times 3 \text{ ton/ac/yr} \times 4 \text{ times removal rate} = 108 \text{ lb } P_2O_5/\text{ac/yr}$$

If a pound of phosphate costs $0.25 then this will cost $27 per acre per year or $9 per ton of hay for phosphate when weather and other management conditions allow a 3-ton yield. Such an application rate does not need to be continued for many years. If using corrective fertilization at high rates, yearly soil testing may be desired. A practical lower cost alternative is to apply phosphate at two times the removal rate, accepting a little lower yield the first few years as the soil fertility increases and soil testing every 3 years.

If this field has a high K soil test value we may not need K fertilization. If the soil test is low for K and for soil pH, these additional costs must be added to determine the cost of improving the site and long-term hay or pasture production. This type of exercise shows the high cost for "improving" cheap abandoned land and bringing it back into production.

VALUE OF FERTILIZER NUTRIENTS

The value of fertilizer nutrients can be calculated from the price of the fertilizer material and the nutrient content in the fertilizer. When using single-element fertilizers such as urea this is simple. For multinutrient fertilizers such as diammonia phosphate or liquid solutions, it is more difficult. Because the single-nutrient fertilizers are often the least expensive and are the bases of mixed dry fertilizers, they do serve as a baseline for nutrient values. If the total value of a mixed fertilizer or solution is lower than the sum of individual nutrients, than it is the best buy.

Using urea as an example, let's calculate the cost per pound of N. Let's say a ton of urea costs $240. This is a cost of $0.12 per pound of fertilizer ($240 per ton/2,000 pounds per ton). Because urea has a label ratio of 46-0-0, we know that there are 46 pounds of N per 100 pounds of fertilizer or 0.46 pound of N per pound of urea. So the cost of N from urea is $0.261 per pound ($0.12 per pound/0.46 pounds N per pound of fertilizer).

The same process can be used with the cost of triple superphosphate (0-46-0) to calculate the value of P_2O_5 and the cost of muriate of potash (a.k.a. potassium chloride, KCl) (0-0-60) to calculate the value of K_2O.

VALUE OF CROP RESPONSE

The value of hay sold or purchased is relatively easy to assess. However, for pastured crops it is easier to look at the pastured crops' replacement value compared to a harvested feed that would be fed if the pasture were not available. In areas that have an established market for pasture, the market value of an animal unit grazing day can be used.

SUMMARY

To minimize fertilization costs, a good soil testing program can be used to determine soil fertility and track the effectiveness of nutrient cycling on the farm. The major volume and value of nutrients on a livestock farm are those recycled in the pasture and hay meadow rather than those leaving the farm in animals, milk, or wool. On fields having a low soil test index for a nutrient or pH, corrective fertilization and liming is necessary to obtain optimum forage yields as determined by the soil RYE and local economic conditions. Where crops are removed from the field, returning nutrients in proportion to their removal rate is necessary to prevent reducing soil fertility. Returning manure to the fields where the feed came from is practical nutrient recycling on the farm. Using these tools and techniques, the manager can develop a fertility program best suited for the operation.

CHAPTER 4
Nutrient Management in Forage-Livestock Systems

Paul R. Peterson, James B. Cropper, Edward B. Rayburn, and William L. Stout, Jr.

Understanding and managing the cycling of nutrients in forage-livestock systems is critical to optimizing livestock production and its environmental impact. Managing the nutrient cycle in a pasture is very different from managing nutrients when forage is harvested and fed to animals in confinement. Nutrient cycling in pastures is a dynamic process involving the natural movement of minerals through soil, plants, and livestock, and the manager's manipulation of these components. In pastures, livestock affect nutrient cycling when they graze pasture plants and consume nutrients, including nitrogen (N), phosphorus (P), and potassium (K), and then return most of those minerals to the soil in their manure (dung and urine).

This chapter focuses on the livestock's and manager's manipulation of components of pasture nutrient cycling. It includes discussion of:

- nutrient flow on and off the farm,
- the characteristics and impacts of livestock manure on pasture,
- how animal behavior relates to nutrient excretion patterns,
- how we can design grazing systems to optimize the distribution of manure return by grazing livestock, and
- applying organic nutrient sources, including animal wastes and biosolids, to pastures.

NUTRIENT FLOW ON AND OFF THE FARM

Only small proportions of the N, P, and K that grazing animals consume are actually retained in animal products (table 4-1, p. 52). Table 4-2 (p. 52) illustrates the nutrient removal in different forage-livestock systems (18). For a beef cow-calf operation stocked at 3 acres per pair per year, the nutrients removed per cow-calf pair are only about 3 pounds N, 1 pound P, and 0.2 pound K per acre per year. Dairy cows remove more nutrients. For a pasture-based dairy stocked at 3 acres per cow per year where no concentrate is fed, net nutrient removal is about 30 pounds N, 5 pounds P, and 8 pounds K per acre per year. When purchased concentrates are fed, there is a net import of 148 pounds N, 32 pounds P, and 94 pounds K per cow. Thus, a dairy farm where purchased concentrates are fed can actually have increasing nutrient levels on pastures.

Table 4-3 (p. 53) lists the animal nutrient and plant fertilizer content of common forages, feeds, and supplements. Purchased feeds can be a major source of nutrients entering the farm. Purchased grain can bring in large amounts of P and N that, if not managed properly, can increase the risk of nutrient loss to the environment. Lactating dairy cows may excrete 65% of the N they consume in supplemental feed onto pasture (31). Purchased hay and silage contain high levels of K, and if not managed properly, can cause animal health risks if soil K levels become too high, causing forage crops to have excess K content.

Table 4-1. Elemental mineral composition of farm animal products.					
	N (%)	P (%)	K (%)	Ca (%)	Mg (%)
Beef steer					
Birth	2.62	0.68	0.22	1.25	0.04
250 lb	3.18	0.80	0.24	1.36	0.04
450 lb	3.15	0.65	0.22	1.11	0.04
1,110 lb	2.63	0.79	0.20	1.47	0.04
Cow's milk	3.20	0.09	0.13	0.11	0.01
Lamb, fat		0.49	0.14	0.92	
Sheep, lean		0.52	0.14	0.94	
Sheep, fat		0.45	0.12	0.85	
Wool, raw		0.03	4.67	0.13	

Sources: Some values calculated from data presented in:
Morrison. 1956. Feeds and Feeding. *p. 640. Morrison Publ. Co., Ithaca, NY.*

National Research Council (NRC). 1988. Nutrient Requirements of Dairy Cattle. *pp. 99–100. National Academy Press, Washington, D.C.*

NRC. 1996. Nutrient Requirements of Beef Cattle. *p. 23. National Academy Press, Washington, D.C.*

Reid et al. 1955. Journal of Dairy Science *38:1344.*

Table 4-2. Estimated annual removal of N, P, and K from different forage-livestock management systems			
	N	P	K
Hay	– 50 lb/ton	– 8–13 lb/ton	– 40 lb/ton
Dairy – pasture-based, no concentrate	– 84 lb/cow	– 15 lb/cow	– 23 lb/cow
Dairy – imported concentrates	+ 148 lb/cow	+ 32 lb/cow	+ 94 lb/cow
Cow/calf	– 10 lb/pair	– 3 lb/pair	– 0.7 lb/pair
Stocker	– 10 lb/calf	– 3 lb/calf	– 0.7 lb/calf
Dry cow	0	0	0

Source: Adapted from Lory, J., and R. Kallenbach. 1999. Soil fertility management and nutrient cycling. pp. 73–80. In: J. Gerrish and C. Roberts (ed.). Missouri Grazing Manual. *MU-Extension, University of Missouri.*

Table 4-3. Animal feed nutrient and fertilizer nutrient content of forage and grain crops.

	Animal feed nutrient content % dry matter					Fertilizer nutrient content Pounds/ton at 90% dry matter				
	CP[a]	P	K	Ca	Mg	N	P_2O_5	K_2O	$CaCO_3$	$MgCO_3$
Pastures										
Pasture, grass	20	0.38	3.38	0.43	0.22	58	16	73	19	14
Pasture, mixed mostly grass	22	0.38	2.76	0.75	0.26	63	16	60	34	16
Pasture, mixed mostly legume	22	0.35	2.65	0.99	0.29	63	14	57	45	18
Pasture, legume	24	0.33	3.07	1.21	0.30	69	14	67	54	19
SD[b]	4	0.08	0.71	0.22	0.05	12	3	15	10	3
Hay crops										
Hay, grass	11	0.21	1.76	0.52	0.19	30	9	38	23	12
Hay, mixed mostly grass	12	0.23	1.86	0.69	0.21	35	9	40	31	13
Hay, mixed mostly legume	17	0.26	2.16	1.10	0.25	48	11	47	49	16
Hay, legume	19	0.25	2.33	1.39	0.28	56	10	51	62	17
Hay, small grain	9	0.21	1.72	0.40	0.15	26	9	37	18	9
Straw	6	0.11	1.50	0.28	0.12	17	5	33	13	7
Corn stalks	6	0.13	0.90	0.40	0.15	6	2	7	7	3
SD	3	0.05	0.50	0.30	0.05	9	2	11	13	3
Silage crops						**Pounds/ton at 33% dry matter**				
Silage, corn	9	0.21	1.05	0.22	0.16	10	3	8	4	4
Silage, grass	14	0.28	2.25	0.62	0.21	15	4	18	10	5
Silage, mixed mostly grass	14	0.27	2.17	0.80	0.22	15	4	17	13	5
Silage, mixed mostly legume	18	0.29	2.37	1.13	0.25	19	4	19	19	6
Silage, legume	20	0.29	2.48	1.27	0.26	21	4	20	21	6
Silage, small grain	13	0.30	2.40	0.50	0.19	14	5	19	8	4
Silage, sg and leg	17	0.34	2.88	0.78	0.22	18	5	23	13	5
SD	3	0.05	0.60	0.26	0.05	3	1	5	4	1
Grains						**Pounds/bushel at 85% dry matter**				
Corn, shell	10	0.30	0.47	0.03	0.12	0.77	0.33	0.27	0.04	0.20
Wheat	14	0.41	0.50	0.13	0.16	1.14	0.48	0.31	0.17	0.28
Oats	13	0.38	0.55	0.11	0.15	0.58	0.24	0.18	0.07	0.14
Barley	13	0.39	0.54	0.08	0.14	0.85	0.36	0.27	0.08	0.20
Soybeans	41	0.59	1.91	0.25	0.25	3.35	0.69	1.17	0.32	0.44
SD	2	0.09	0.10	0.02	0.04	0.15	0.10	0.06	0.02	0.07

[c]CP = 6.25 N P_2O_5 = 2.291 P K_2O = 1.205 K $CaCO_3$ = 2.497 Ca $MgCO_3$ = 3.467 Mg

[a]*CP-crude protein.*
[b]*SD-standard deviation; ±1 SD around the mean includes 66% of the observations that make the mean and represents the effects of management, weather, and plant maturity on the forage quality of samples submitted for analysis.*
[c]*Conversion equations for calculating crude protein, oxide, or carbonate equivalent for plant nutrients.*
Source: Adapted from the 1994 Annual Summary of Forage and Grain Analysis, Dairy One Forage Testing Lab, Ithaca, NY.

Table 4-4 provides an example of a farm having 6 months of grazing and 6 months of hay feeding for a beef operation or pasture-based dairy. The highest volume and value of nutrients are those nutrients recycled in the pasture. Pasture and hay together represent more than $135 worth of nutrients recycled. The beef system transports less than $8 worth of nutrients off the farm, but the dairy system removes about $25 worth of nutrients.

Because livestock excrete most of the minerals they consume, manure from grazing animals represents a valuable source of fertilizer nutrients if it is distributed uniformly across the pasture. Cattle defecate and urinate, on average, 12 and 8 times per day, respectively, but this can vary substantially (2, 26, 41, 43). Sheep urinate 18–20 times and defecate 7–26 times per day (14, 31). Since grazing animals generally excrete at sites other than where the nutrients in that excreta are consumed, nutrient redistribution occurs. The manager should strive to design and manage a grazing system to enhance uniformity of manure and thus nutrient return without incurring excessive costs or unduly affecting animal performance.

Table 4-4. Magnitude of fertilizer nutrients that are recycled in pasture when grazed for 6 animal unit months (AUM, a 1000-lb animal grazing for 1 month) and hay compared to the magnitude of fertilizer nutrients that are removed from a farm in feeder steers or milk sold.

	N	P_2O_5	K_2O	$CaCO_3$	$MgCO_3$	
			Pounds Nutrients Recycled			
Pasture 6 AUM (2.3 tons dry matter (t DM))	161.0	39.8	152.2	85.7	41.2	
			Pounds Nutrients Removed			
Hay 6 AUM (2.3 t DM)	89.1	24.2	103.1	79.3	33.5	
550 lb steer	17.0	7.4	1.3	13.9	0.8	
10,000 lb milk	51.2	22.9	16.9	30.0	3.5	
			Value Nutrients Recycled ($)			**Total**
Pasture 6 AUM (2.3 t DM)	48.84	10.78	24.09	1.06	0.57	85.34
			Value Nutrients Removed ($)			
Hay 6 AUM (2.3 t DM)	27.01	6.56	16.32	0.98	0.46	51.33
550 lb steer	5.15	2.02	0.21	0.17	0.01	7.55
10,000 lb milk	15.53	6.20	2.67	0.37	0.05	24.82
			Fertilizer and Nutrient Value Used			
Fertilizer	**$/ton**		**$/lb nutrient**		**Nutrient**	
Urea	279		0.303		N	
Triple superphosphate	249		0.271		P_2O_5	
Potassium chloride	190		0.158		K_2O	
Dolomitic lime	26		0.0124 0.0138		$CaCO_3$ $MgCO_3$	

IMPACT OF LIVESTOCK EXCREMENT ON PASTURE PRODUCTION

The response of pasture production to animal excretion includes short- and long-term effects, depending upon which nutrient is considered. Short-term forage production responses are largely due to the N concentration in urine. This short-term response is especially evident in N-deficient pastures. The P and K in livestock feces are the primary nutrients implicated in more long-term pasture responses.

Urine carries 50–80% of the N that is excreted by livestock (42). However, in contrast to the N content of feces, urinary N output varies considerably as N and energy in the animal's diet vary (31). Steers grazing high-quality pasture excreted averages of 0.24 pound N per day in urine and 0.11 pound N per day in feces (3). The area covered by an individual cow's urination ranges from 1.5 to more than 4 square feet (6, 26, 43), but the area of influence of the nutrients in the urine can be considerably larger (7). Average volume of each urination is about 2 quarts for cattle and about 0.15 quart for sheep (14), but this varies greatly among animals and with time of day and water intake. Urine-N is "applied" at rates equivalent to up to 1,100 pounds N per acre by dairy cows (31). Unfortunately, these high rates exceed the capacity of the soil to absorb and retain the urea-N, resulting in loss of ammonia by volatilization in the short term and by leaching of nitrate-N or denitrification in the longer term.

Cattle defecations range in size from 0.5 to 1.5 square feet ground coverage (6, 26, 43), and each defecation weighs between about 3 and 6 pounds (14). Typical application rates in a dung patch are 930 pounds N per acre, most of which is in organic forms (14). About 22% of N in cattle dung becomes available as inorganic soil N during the season (32).

There are varying claims regarding the percentage of the pasture area that falls within the area of influence of livestock excretions, and thus varying theories about the influence of livestock excrement on pasture productivity. Estimates of the percentage of pasture area covered by at least one excretion per year range from about 10 to 35% (6, 14, 24, 41). Stocking rate, stocking method, forage quality, season, and climate all interact to influence the percentage of pasture area covered and influenced by livestock excrement.

New Zealand researchers reported that cattle manure piles affect forage growth in a zone fivefold larger than the area covered by the pile (7). They found that the area influenced had 14% more forage yield and 23% more P than pasture areas outside the area of influence. The large zone of influence was attributed to lateral wash by rain and lateral spread of roots and stolons (e.g., white clover). Based on their results, they estimated that under heavy stocking, more than 50% of a pasture is directly influenced by manure pile nutrients at any one point in time. They also speculated that this is an underestimate, because in a grazier's pasture, lush spring and fall pasture usually results in more liquid and widely spread manure, and because cattle kick and carry manure on their feet as they travel about a pasture. Data from Missouri suggest that with frequent rotation, 40% of the pasture area might be within the zone of influence of a manure pile from the same year (18).

Bermudagrass receiving 180 pounds N, 38 pounds P, and 45 pounds K per acre per year and growing within the confines of livestock excretions produced 58 and 263% more forage in areas influenced by dung and urine, respectively, during 5 months of grazing than did control areas receiving only synthetic fertilizer (6). Given about 35% total coverage by either manure or urine, this translated into 14% more forage production per acre. However, these scientists did not quantify forage yield benefits within a potential zone of influence outside of the area directly under an excretion, so 14% is likely an underestimate of the boost in productivity due to excretions. Research in the United Kingdom documented that excretal return by sheep increased perennial ryegrass-white clover pasture yield by 26 and 53% when stocked at 10 and 20 yearling wethers per acre, respectively (5).

The annual production of N-fertilized orchardgrass pasture in a 6.7-square-foot area surrounding dairy cow feces and urine was increased by about 15 and 55%, respectively, in Pennsylvania (37) over pasture receiving only the N fertilizer. These values may be artificially high because they represent average yield response over 3 years to manure or urine applied on the same spot in each of 3 years. However, these orchardgrass pastures were receiving from 175 to 250 pounds N per acre per year; thus, a significant response to additional N (urine) would not have been expected.

Despite clear benefits of livestock excrement to pasture productivity, there can also be significant uncontrolled losses of N from these excretions, especially from urine. Growing plants recover relatively small proportions of urine-N due to N losses and N immobilization by microbes. Recovery ranges from 20 to 30% but is better in spring and summer because of more rapid plant growth (31). Thus, pastures can be quite deficient in N because up to 50% or more of the N applied by the grazing animal can be lost via a combination of volatilization, denitrification, and leaching. This is particularly true of pastures without a significant legume component. On average, 25% of the N in urine and feces deposited on a pasture can be lost to the atmosphere via ammonia volatilization (18). Nitrate leaching can be high where N fertilization rates are high (31), representing up to 60% of annual fertilizer application (16). New Zealand studies have indicated that even grass-legume mixtures receiving no N can have losses of N via leaching (11, 36), but more deeply rooted legume-grass mixtures, such as alfalfa, may have less N leaching (31). However, 43% of the 450 pounds N per acre equivalent applied in urine was taken up by perennial ryegrass-white clover, and only 8% of the N leached (12). Leaching losses are most probable during autumn to spring when pasture growth is slow and precipitation exceeds evapotranspiration (31). N fertilization rates on pasture should be reduced near or during times of minimal plant growth.

Russelle (31) suggested the following strategies to improve N use efficiency:

- Use moderate fertilizer N rates (less than recommended for mown forages) and divide the total into small split applications.
- Avoid fertilizer application during or immediately after drought periods.
- Include moderate populations of legumes to provide symbiotically fixed N.
- Select supplemental feeds high in rumen-degradable carbohydrates that improve N use efficiency of the animal but that do not increase N loading on the pasture.
- Adjust stocking rates to match feed availability in herbage and supplements.
- Avoid grazing in late autumn to reduce overwinter leaching losses of nitrate from urine spots.
- Install drain tiles only on those sites where substantial plant yield gains will accrue.

SPATIAL DISTRIBUTION OF NUTRIENTS ON PASTURES

If livestock spent proportionate amounts of time on all areas of a pasture, excretions would uniformly cover the pasture and optimize nutrient cycling. This is not the case, however, because livestock are attracted to certain areas of pastures.

Manure deposition is influenced primarily by water, shade, and topography. Location of water has a profound effect. Iowa researchers reported that after

5 years of continuous grazing, soil in the 5% of the pasture area closest to water had P levels sixfold greater than the remainder of the pasture and K levels almost fourfold greater (40). Missouri researchers found that after 4 years of a three-paddock rotation on a 32-acre pasture stocked at 2.6 acres per cow-calf pair, soil in a zone extending 40 feet from the watering site had P and K levels 4 and 10 times those, respectively, in the rest of the paddock (13). Both research groups attributed these profound gradients in nutrient levels to higher rates of manure deposition closer to the watering locations. The Missouri researchers also noted a greater gradient toward water in paddocks that were long and narrow. This could be due to the length-width ratio itself or to the distance cattle were required to travel to reach water.

In Georgia, tall fescue pastures that were continuously grazed by steers had higher concentrations of P and K 3 feet from permanent shade and water sources, but P and K levels at 30 feet did not differ from the remainder of the pasture area (33). However, their observations were on experimental-scale pastures less than 2 acres in size with maximum travel distances to water and shade of 400 feet. On a commercial scale in the Northeast, the zone of P and K accumulation around shade and water is undoubtedly larger.

The concern with this gradient of nutrients toward water is twofold. First, an environmental problem can develop at watering sites. Second, the nutrients building up at the watering site are being pulled from the bulk of the pasture area. Thus, the majority of the pasture becomes depleted of these nutrients.

Pasture systems requiring animals to travel a lane to water result in a loss of excretal nutrients off the pasture. At an experimental scale (32-acre pasture system), losses averaged 15% but were as high as 25% during August (27). (Cows spend more time going to water in August so more manure was dropped on the lane.) At a commercial scale, manure loss off of pasture areas to lanes could be larger. Lanes that provide water access often do not provide much forage. Thus, the manure dropped in the lane represents a loss of nutrients off the pasture and a potential environmental problem.

Shade is a primary location for livestock camping and loafing, especially during hot, sunny days. However, cattle often camp under shade trees even during cloudy and/or cool days. Time spent camping under shade results in accumulation of manure and urine, and thus P and K, at the shade source. Missouri researchers found that the gradient of P and K concentration toward a single shade tree increased more sharply and to a greater extent than that toward the watering site (13, 18).

Endophyte-infected tall fescue also plays a role in nutrient distribution. The gradient toward shade is even worse on endophyte-infected tall fescue pasture because cattle are more attracted to shade when stressed (33, 44). In Georgia, P levels 3 feet from shade and water were 64% higher on continuously grazed endophyte-infected vs. low-endophyte tall fescue pasture, but endophyte status did not affect K or Mg accumulation (33). This difference would likely be less pronounced in the Northeast due to cooler temperatures and less tall fescue.

The need for shade on pasture for optimization of livestock comfort and performance is a controversial subject (4). In the Northeast, there are undoubtedly several days during the summer when beef cattle "need" shade, and several weeks when dairy cattle "need" shade. Thus, it is probably a good plan to have shade available in a few paddocks. However, for most of the paddocks on the farm, unless shade is moveable or fairly uniformly distributed throughout the paddock, it is probably best not to intentionally provide shade. In fact, consider removing or fencing off shade trees in most paddocks to negate the powerful influence they have on animal behavior and thus the uniformity of excretal return. However, access to shade may be required for organic certification.

Topography also influences patterns of excretal return. Sheep and cattle generally prefer the highest ground available to them to camp, especially during wet or warm weather, and thus deposit disproportionately more manure there. On border-strip irrigated pasture, sheep deposited 34% of their manure on the 15% of the land area in levees (45). The tendency for higher ground to be drier and have greater exposure to cooling breezes is probably the reason for this.

GRAZING MANAGEMENT FOR NUTRIENT MANAGEMENT

Rotational stocking is an effective tool to improve the uniformity of distribution of livestock excreta (28). Research in New Zealand (21), the United Kingdom (30), and the United States (27, 41) has documented that increased stocking densities improve uniformity of excretal return by both cattle and sheep. Stocking density is the number of animals present per unit land area at a given point in time. Since rotational stocking involves the subdivision of larger land areas into smaller subunits, stocking densities are inherently increased. Also, rotational stocking "forces" animals to spend roughly equal amounts of time on the entire pasture system area by eliminating their ability to maintain permanent camping and loafing sites.

In an experiment with sheep in Australia, rotational stocking reduced the percentage of manure deposition in 1% of the pasture area from 24 to 6% (15). A decrease from 46 to 32% was observed on 10% of the area. In a 24-paddock rotational system in Oklahoma, 35% of the paddock area was covered by excretions in one grazing season (6). Missouri research indicated that rotational frequency has a profound effect on the length of time required for every square yard of pasture to receive at least one manure pile (18, 27), ranging from a low of 2 years for a 2-day rotational frequency to a high of 27 years under continuous stocking. Thus, the main grazing area in a continuously grazed pasture has a nutrient cycle more like a hay field. Petersen et al. (26) calculated that after 5 years, 75% of a pasture would be covered by at least one excretion.

Providing water in paddocks will minimize loss of excreted nutrients off pasture. However, even with this strategy, animals will defecate and urinate more near the water tank during hot weather (41). Portable water sources may help to distribute manure more uniformly during hot weather.

Having multiple watering sites no greater than a quarter mile away from each other is helpful. If possible, place water troughs near the center of pastures rather than against a fenceline and/or in the paddock corner. Pastures with streams should have alternative water sources. Where alternative off-stream waterers are provided, animal time in the stream is dramatically reduced (19). Livestock water ponds should be fenced and the water piped to troughs. Placing troughs, mineral feeders, and other inducements at or near the top of slopes in steeply sloping pasture will provide incentive for animals to graze and manure the steep areas. Poorly laid out pasture subdivisions served by single-source water, feeding, salt, mineral, and shade areas will cause livestock to camp at these sites just as they do on continuously stocked pastures.

Where water is provided via lane access, some strategies can be implemented to reduce this loss (6). First, provide nothing but water at the watering location, that is, do not intentionally place water in shade and do not provide mineral or supplement at the watering location. Second, keep the lane relatively narrow to discourage the entire herd from lingering. Third, consider gravelling the watering area and the lane. Cattle will walk on gravel to access water, but are not comfortable standing or laying on gravel for extended periods.

Good grazing management usually results in higher quality, more succulent forage, which in turn leads to more liquid, spread out, rapidly degraded manure. Dragging often becomes necessary only when forage quality is such that manure piles are more solid. Dragging is a common technique in which a chain drag is used to spread manure.

Dalrymple et al. (6) provided a good summary list of strategies to favor even distribution of livestock excrement on pasture:

- Keep salt and minerals in a portable feeder on the paddocks, place it where cattle otherwise would not be attracted, and change its location every grazing period.
- If providing supplemental feed, offer it on the ground under an electric wire and vary the location at each feeding.
- Do not provide shade on purpose. If shade is present, conduct all other activities elsewhere in the paddock.
- Increase stocking density as much as possible. This can be accomplished by increasing paddock numbers and reducing grazing period length (i.e., rotational stocking).

APPLYING ANIMAL WASTES AND BIOSOLIDS TO PASTURES

Since most nutrients are recycled within a pasture, pastures should not be viewed as waste disposal areas of endless capacity to absorb and use nutrients from confined animal operations. Low amounts of nutrients leave in the animal products harvested from the pasture. Thus, even with nonuniform excretal distribution and loss of some nutrients, application of organic materials like manures, sludge, or biosolids can lead to overapplication of some nutrients, especially if these materials are used to supply the entire N rate needed to optimize forage yield and quality. They are rather complete fertilizers containing many elements.

Table 4-5 (p. 60) illustrates the wide variation in composition of animal manures. In general, poultry manure has the highest composition of N, P, and K, and dairy cattle the least. The ratios among N, P, and K content vary significantly as well, with dairy cattle manure having the highest N:P ratio and poultry litter the lowest (8, 34).

Animal manures vary in how rapidly organic N is mineralized during the first year after application. On average, 35, 50, and 60% of organic N becomes plant-available during the first year for cattle, swine, and poultry manure, respectively. In subsequent years, all animal manure types generally have similar rates of mineralization, averaging about 12 and 5% in years 2 and 3, respectively (39). Of manure applied to pasture, about 20% of the N can be lost to the atmosphere in the first 6 hours after spreading manure, and up to 60% can be lost over several days (38).

Some Virginia data demonstrate the tremendous potential yield response of tall fescue in response to organic sources of N (table 4-6, p. 61). Organic sources of N were surface-applied in March at rates designed to supply 250 pounds N per acre during the year of application; however, results revealed that N mineralization rates were different from the book values used to determine application rates. Organic N sources had high N uptake, which boosted tall fescue's crude protein (CP) content. Fiber concentrations were unaffected by any treatment and averaged about 30% acid detergent factor and 60% neutral detergent factor (10).

Manure application to pasture must be done with prudence. Heavy applications of manures containing N and K may cause seasonally low dietary levels of Mg in the forage ingested by livestock, just as chemical fertilizers can, leading to potential metabolic problems such as grass tetany. If poultry litter is used as a nutrient source, prudent rates must be applied, preferably after periods of

Type of animal	N	P %, "as-is" basis	K	Approx. N:P:K
Dairy cattle				
Lactating cow	0.56	0.09	0.33	6:1:4
Dry cow	0.44	0.06	0.28	7:1:5
Heifer	0.36	0.05	0.28	7:1:6
Beef cattle				
Cow	0.52	0.19	0.41	3:1:2
450–700 lb	0.52	0.17	0.34	3:1:2
750–1,100 lb, high forage diet	0.52	0.19	0.41	3:1:2
750–1,100 lb, high energy diet	0.59	0.18	0.41	3:1:2
Swine				
Nursing, 0–40 lb	0.57	0.24	0.33	2:1:1
Grower	0.66	0.25	0.35	3:1:1
Replacement gilt	0.73	0.24	0.40	3:1:2
Gestating sow	0.70	0.23	0.45	3:1:2
Lactating sow	0.78	0.25	0.50	3:1:2
Boar	0.73	0.24	0.49	3:1:2
Poultry				
Layer	1.77	1.15	1.25	1:1:1
Broiler	1.94	0.97	1.14	2:1:1
Turkey	3.62	1.65	1.85	2:1:1
Horse	0.56	0.10	0.38	6:1:4

Source: Edwards, D. R. 1996. Recycling livestock wastes on pastures. pp. 45–63. In: *R. E. Joost and C. A. Roberts (ed.).* Nutrient Cycling in Forage Systems. *Columbia, MO. 7–8 March 1996. Potash and Phosphate Institute, Manhattan, KS.*

Adapted from Soil Conservation Service. 1992. Agricultural Waste Management Field Handbook. *U.S. Dept. Agric., Washington, D.C.*

Table 4-5. N, P, and K concentrations of selected animal manures "as excreted."

high spring grass growth rate. Poultry litter is particularly high in N and K. Keeping poultry litter spreading rates at agronomically appropriate levels will help avoid grass tetany outbreaks in pastured livestock. High rates of poultry litter applied to endophyte-infected tall fescue pastures can also intensify bovine fat necrosis outbreaks. No more than 4 tons per acre of poultry litter should be spread on tall fescue pastures (1).

Sheep are susceptible to copper (Cu) toxicity, so caution should be exercised before stocking sheep on pastures where recent applications of poultry litter or swine manure were made. Both manures may contain high Cu concentrations because Cu salts are often fed as wormers to both these livestock types. It's best to determine the Cu levels in the manure as a precaution.

Table 4-6. Total season yield (ton DM/acre) and season average forage quality (% DM) of tall fescue harvested four times (May, June, August, and October) during 1 year in Virginia.

N source/amount	Total N applied (lb N/ac)	DM yield (t/ac)	N uptake (lb/ac)	% CP
0 lb N/ac (control)	0	2.5	95	11.8
100 lb N/ac	100	5.0	191	12.0
200 lb N/ac	200	6.0	275	14.4
300 lb N/ac[a]	300	7.4	363	15.5
400 lb N/ac[a]	400	8.0	368	14.3
Lime-stabilized biosolids[b]	707	7.9	381	15.0
Anaerobically digested biosolids	1,025	8.9	457	16.1
Poultry litter	954	8.3	427	16.2
Least significant difference (LSD) (0.05)		1.2		1.1

[a] *The 300 and 400 lb/ac N rates were applied in split applications of 200+100 and 200+100+100, respectively.*
[b] *Organic sources were applied in March.*

Source: Evanylo, G., and P. Peterson. 2000. Availability of N in Biosolids for Tall Grass Hay Production. *Final report to the T. M. Hepler Endowment Committee, Virginia Tech. 9 pp.*

Biosolids are solid, semisolid, or liquid materials, resulting from treatment of domestic sewage, that have been sufficiently processed to permit their safe land application. Biosolids are produced primarily through biological treatment of domestic wastewater, and are sanitized to control disease-causing organisms. Biosolids contain mineral and organic matter. The mineral matter includes plant nutrients, and organic matter is a source of slow-release nutrients and soil conditioners (9). Concentrations of nutrients in biosolids can vary significantly (table 4-7); thus, the actual material being considered for pasture application should be analyzed. Trace elements are found in low concentrations in biosolids. The trace elements in biosolids are "heavy metals"; federal and state regulations exist for a number of these (9). Correcting soil pH and applying biosolids to soil with a high cation exchange capacity will help to minimize problems with heavy metals.

Table 4-7. Typical nutrient concentrations (% in dried solids) in biosolids from all processes.

Nutrient	Range (% in dried solids)	Mean (% in dried solids)
Total N	<0.1–17.6	3.9
NH_4-N	0.0005–6.7	0.65
NO_3-N	0.0002–0.49	0.05
Total P	<0.1–14.3	2.5
K	0.02–2.64	0.40
Ca	0.1–25.0	4.9

Source: Evanylo, G. K. 1999. Land Application of Biosolids for Agricultural Purposes in Virginia. *Virginia Cooperative Extension Publ. No. 452-300.*

Adapted from Sommers, L. E. 1977. Chemical composition of sewage sludges and analysis of their potential use as fertilizers.

Evanylo (9) summarized the steps in determining biosolid application rates on agricultural land:

1) Determine nutrient needs for expected forage yield and soil test levels.
2) Calculate biosolids agronomic rates based on crop N needs, soil test P needs, or soil lime requirement.
3) Calculate supplemental fertilizer needs by subtracting the amount of plant-available N, P, and K supplied by biosolids from the crop N, P, and K needs. Plant-available N is the actual amount of N in the biosolids that is available to the crop during the season of application. Suggested values for plant-available N of different biosolids are shown in table 4-8. The U.S. Environmental Protection Agency estimates that 50% of the P and 100% of the K applied in biosolids is available for forage uptake in the year of application.

Regulations require that bulk biosolids be applied at a rate that is equal to or less than the agronomic N rate for the forage and soil type. The relative concentrations of nutrients in biosolids are rarely present in the proportions required; thus, supplemental fertilization may be needed. In particular, biosolids often have low concentrations of K. Biosolids normally supply similar amounts of plant-available N and P, so the potential for excess P exists if application rates are based solely on supplying N needs. Applying biosolids at rates to supply P eliminates this concern; however, additional N would likely be needed on grass pasture (9).

Biosolids are generally sprayed or spread and left on the surface of pastures. Split applications are recommended for rates of liquid biosolids exceeding 2–3 dry tons per acre. Treated and composted municipal sludge can be spread on pastures just before spring green-up or on recently grazed rotational paddocks before forage regrowth occurs.

Livestock grazing pastures treated with sludge can ingest contaminants contained in the sludge by eating soil, fouled leaves and stems, or plants that have taken up the contaminants through their roots and have metabolized them into their tissues. The two nonessential elements of primary concern in municipal sludge are lead (Pb) and cadmium (Cd). At the levels commonly found in soils or sludges, these two elements have no detrimental effect on plant growth. However, they can cause serious health problems to livestock that eat plants or soil that is sufficiently contaminated with them. Sludge can also contain high levels of Cu and zinc (Zn). These trace elements can become toxic to plant growth if soil concentrations are excessive, but toxicities are extremely rare (17).

Table 4-8. Estimated percentage of organic N that becomes available for plant uptake at various times after application for various biosolids.

Time after application	Lime stabilized	Aerobically digested	Anaerobically digested	Composted
Years	Plant-available portion of organic N (%)			
0–1	30	30	20	10
1–2	15	15	10	5
2–3	7	8	5	3

Source: Evanylo, G. K. 1999. Land Application of Biosolids for Agricultural Purposes in Virginia. *Virginia Cooperative Extension Publ. No. 452-300.*

If composted sludge is applied on pasture at rates greater than 30 tons per acre, cattle and sheep ingesting the compost can become Cu deficient. This seeming contradiction occurs because molybdenum is often found in high concentrations in municipal waste sludges, and it can bind with Cu and hinder its absorption in animals consuming these forages (17).

All of these heavy metals have maximum soil loading rates that must be observed by the receiver of biosolids for land application. Once those levels are reached, sludge applications must cease. Annual soil tests for these elements are needed to ensure that they will not accumulate above the maximum levels. Check with your state environmental agency for their regulations on land application of sludge. Guidelines can vary from state to state. Applications are also restricted based on ground slope, amount of vegetative cover, and distance to water.

SUMMARY

Since livestock excrete most of the fertilizer nutrients they consume from pasture forage, grazing system design and management should attempt to enhance uniformity of excretal return to the extent feasible. Livestock tend to excrete most near water and shade and on higher ground because this is where they linger. Increasing stocking density and reducing grazing period length enhances uniformity of manure deposition. These processes can most effectively be accomplished via implementation of rotational stocking systems, with greatest uniformity under rapid rotations. Provision of water in paddocks reduces the loss of excretal nutrients that occurs from pasture when cattle are forced to travel a lane to access water. Animal wastes and biosolids can be effective, safe organic nutrient sources for enhancing pasture production, but should be applied only after and in light of analyses of the soil and material to be applied.

CHAPTER 5
Environmental Impacts of Grazing

William L. Stout, Jr., Andrew N. Sharpley, Lloyd B. Owens, James B. Cropper, and Ronald R. Schnabel

INTRODUCTION

The prevailing perception among many graziers, scientists, and extension personnel is that the practice of grazing is inherently "environmentally friendly" and that grazing is by nature better for the environment than confinement animal production systems. Although the overall effects of grazing on the environment can be positive, this is not necessarily true. Grazing animals are on the landscape consuming and recycling nutrients without the controls on these activities imposed by properly designed and managed confinement animal production systems. Grazing can be "environmentally friendly" only if we understand the impacts that grazing animals have on the land and water and how to manage those impacts to maintain the overall quality of the environment. Thus, the purpose of this chapter is to explain the major environmental impacts of grazing and identify some management options to mitigate them.

SOIL COMPACTION

Soil compaction is the compression of soil when a load is applied to it while it is moist enough to be molded or displaced (87). Soil compaction can increase both soil density and soil strength. These increases, caused by livestock traffic, are most severe in the top 1 inch of soil, but can extend deeper (39). Bulk density changes can extend to 6 inches and soil strength changes to 12 inches into the soil (39).

Two mechanisms of soil compaction occur in pastures. In the first, fine-textured soils become plastic when wetted sufficiently, and soil pore space is reduced when pressure is applied to the soil. The second occurs in saturated soils that have reached the liquid limit and livestock hooves cause soil to be laterally displaced. The displaced soil increases in density when it dries because soil structure was destroyed during lateral displacement (26).

Livestock do not exert undue pressure on the soil when the soil moisture is below field capacity. The pressure they place on the surface soil is comparable to that of properly inflated tractor and haymaking equipment tires (27). Mature sheep, depending on weight and hoof size, exert standing pressures between 10 and 14 pounds per square inch (psi) (75). Mature cattle, depending on weight and hoof size, exert standing pressures between 18 and 26 psi (72). The walking load is about twice as great as the standing load because only two hooves are in contact with the soil at any time (87). Total traffic area (hoof area steps) per animal is 0.008 acre per day for sheep and 0.02 acre per day for dairy cows (52). However, because sheep are normally stocked six times higher than cows, sheep traffic area per animal would be 0.048 acre per day.

The occurrence of soil compaction by livestock traffic is greatest when soils are moist (38), but the resulting increased density near the soil surface remains when the soil is dry (26). Thus, year-round occupancy of pastures where winters are wet is particularly harmful to fine-textured soils (e.g., silt loam to clay loam). In these types of soils, compaction reduces the available water for plant growth because of (i) loss of water due to increased runoff, and (ii) the loss of soil pore space that can hold water at tensions that are low enough to allow plant uptake. Conversely, a sandy loam soil compacted by cattle actually held more water because the soil macropore spaces were reduced, allowing the soil to retain more water.

Soil compaction is second only to plant effects (i.e., surface vegetative cover and rooting density and depth) in influencing infiltration rate and thereby runoff from grazed pastures (2). Half the runoff coming from pastures that are used to overwinter livestock occurs during the dormant season, when compaction is highest. Most of the sediment losses occur in late winter and early spring because the soils are wet, loosened at hoof prints, and often exposed due to close grazing and plant loss (55).

Once compacted, pasture soils tend to remain so until rested from livestock traffic. Frost action, shrink-swell of some clays, plant root growth and decay, and earthworm and other burrowing animal and insect activity can lessen soil compaction with time. Mechanical soil aerators can be used, but the results have been mixed. The primary problem with using soil aerators is the expense involved and the brief time for which the loosening lasts. If livestock have access to the aerated pasture shortly after the work is done and if the sward is not dense and the soils are wet, serious renewed compaction will occur.

Soil compaction in pasture can be reduced by deferring grazing on wet pastures, moving watering and feeding stations, maintaining a dense sward, and leaving as tall a stubble as possible without risking desired plant cover or reducing utilization rates below economic thresholds. Sod-forming forages, such as reed canarygrass, can reduce soil compaction on chronically wet soils.

NUTRIENT IMPACTS

It is well documented that nitrogen (N) and phosphorus (P) can be lost from pastures. Although these losses can be attenuated, they cannot be stopped altogether (40). The pasture manager can do his or her best to keep livestock well dispersed over all pasture acres to avoid creating nutrient hot spots. This is discussed in chapter 4.

Nitrogen

Scientists have identified seven major pathways of N loss from pastures:

- ammonia volatilization,
- denitrification,
- leaching,
- retention in animals or animal products,
- transfer to unproductive areas by animal excretions,
- erosion by wind and water, and
- fire.

In pastured and hayed grasslands the primary loss of nitrogen to the environment on sandy soils (4, 20) is due to leaching; on heavy or poorly drained soils (91) it is due to denitrification.

Nitrate leaching from intensively grazed pastures occurs in humid regions where precipitation exceeds evapotranspiration. Nitrate leaching results from high levels of N fertilization and the uneven recycling of N in urine and dung (61). Decomposing legume nodules after defoliation are also a potential source of nitrate loss (14). Nitrate leached from grazed grasslands far exceeds that leached from similarly fertilized cut grasslands; the leaching rate is affected by soil type and stocking density (24, 77).

In humid temperate regions of the world, an increased stocking rate is the main factor increasing N leaching losses from pasture (4, 15, 76). Nitrate leaching from pastures is characterized by high losses from urine patches superimposed over lower losses from the pasture as a whole (15). The losses from urine patches are affected by animal size, animal type, forage quality and growth stage, and climatic factors (91), while N loss from the pasture as a whole can be affected by soil hydrologic properties, fertilization rate, and pasture species composition (15).

The uneven deposition of urine N in pastures may pose a threat to water quality in many parts of the world (3, 66, 76, 78). Each cattle urine spot affects about a 23.6-inch diameter area (24, 62), and N concentrations under urine spots are equivalent to a 625-pounds-per-acre fertilizer N application. The water contained in the urine facilitates the leaching. In the northeastern United States, urine N does not remain in the soil from year to year but leaches within a relatively short time in the early spring (figure 5-1). In the northeastern United States about 25% of the N excreted in cattle urine can leach below the root zone (79), which can lead to a sizeable amount of nitrate-N leached when projected over the entire pasture (figure 5-2).

Depending on temperature, some of the urea in urine is volatilized as ammonia (28, 29); however, most of the urea rapidly mineralizes to nitrate and is subject to leaching or denitrification (3).

A relatively low cumulative seasonal stocking rate of as little as 81 animal-density (200 animal-density per hectare; mature Holstein dairy cows, 1,300 pounds) per acre would result in a 10 parts per million (ppm) (10 mg per liter) nitrate-N concentration in the leachate beneath an intensively grazed pasture (figure 5-2). This means that while intensive grazing can improve the profitability of small- and medium-sized dairy farms and can provide erosion control, it can have

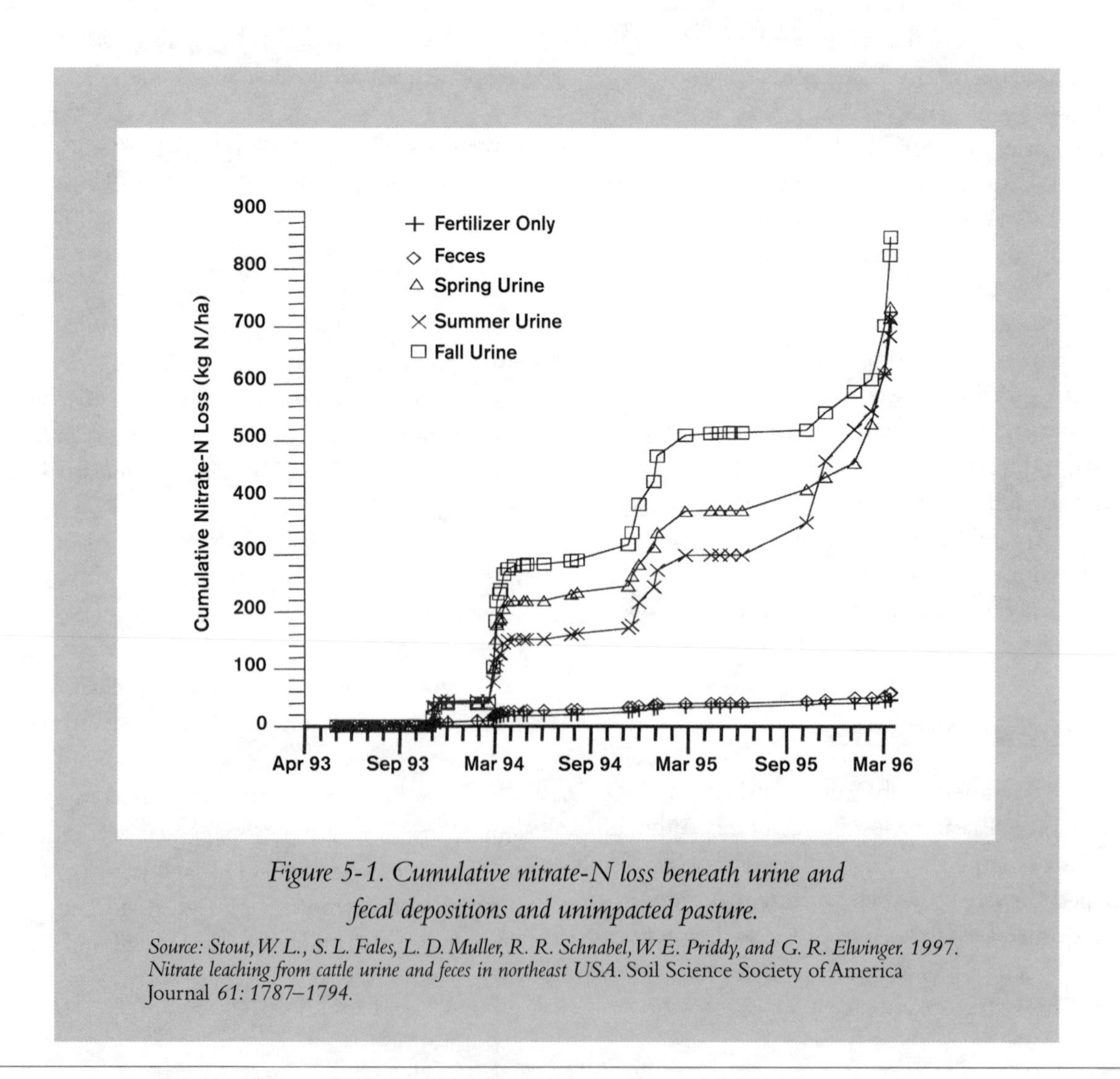

Figure 5-1. Cumulative nitrate-N loss beneath urine and fecal depositions and unimpacted pasture.

Source: Stout, W. L., S. L. Fales, L. D. Muller, R. R. Schnabel, W. E. Priddy, and G. R. Elwinger. 1997. Nitrate leaching from cattle urine and feces in northeast USA. Soil Science Society of America Journal *61: 1787–1794.*

a significant negative impact on water quality beneath pastures.

N losses can be so site-, soil-, climate-, and landscape-specific as to preclude general recommendations regarding overall grazing management to minimize the risk of N leaching. This is particularly true of minimum animal units per acre and N fertility required to reduce the risk of N leaching to acceptable levels. However, research (91) has suggested that 178–223 pounds N per acre or using grass-clover stands with no applied N minimizes N leaching potential under grazing.

Phosphorus

High concentrations of soil P can occur in grazed pastures where manure is concentrated (25, 72). For example, using the Bray-I soil test, researchers (90) have found soil P ranging from 3 to 9 ppm in an ungrazed tall fescue (Kentucky 31) permanent pasture. After 5 years of grazing (2 beef steers per acre), Bray-I P ranged from 10 to 150 ppm, with the higher soil P values extending only 33–66 feet from the water source (89).

The localized accumulations of P where manure is deposited can saturate the P sorption capacity of a

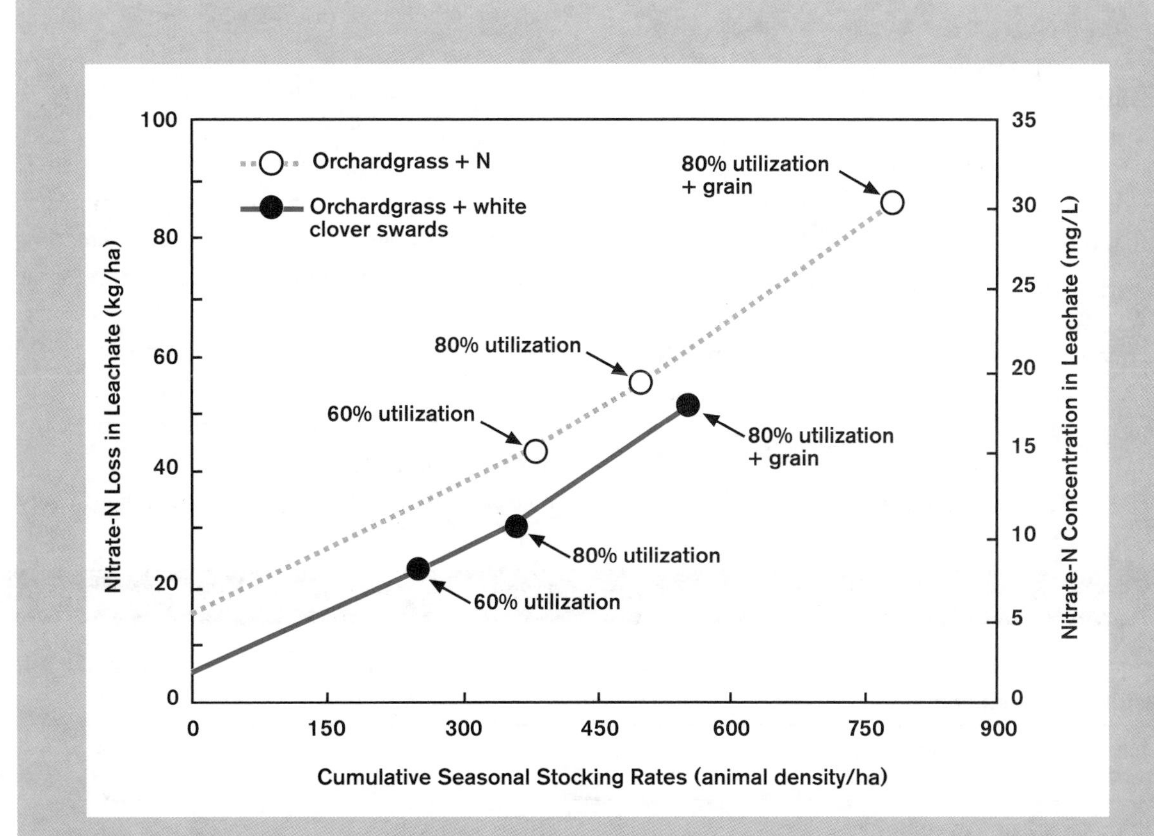

Figure 5-2. Relationship between stocking density and nitrate-N in leachate beneath N-fertilized orchardgrass and orchardgrass/clover. Percent utilization represents how much pasture the animals ate.

Source: Adapted from Stout, W. L., S. L. Fales, L. D. Muller, R. R. Schnabel, W. E. Priddy, and G. R. Elwinger. 1997. Nitrate leaching from cattle urine and feces in northeast USA. Soil Science Society of America Journal *61: 1787–1794.*

soil, increasing the potential for P loss from grazed pastures in runoff or drainage waters (7, 49, 70). For example, researchers (54) have found that decreasing stocking density and duration dramatically reduced runoff and erosion from a pastured watershed in Ohio (table 5-1). Clearly, increased runoff and erosion with grazing will enhance the potential for P loss. In Oklahoma, another research team (53) found that P losses were greater from continuously grazed (4.1 pounds P per acre per year) than rotationally grazed pastures (1.2 pounds P per acre per year). In fact, P losses with continuous grazing were greater than from alfalfa (2.2 pounds P per acre per year) and wheat (2.6 pounds P per acre per year) (52). However, the work of the Ohio scientists (54) shows that when management is changed, the impacts of the previous grazing scheme are not long lasting; they changed within a year.

A third research team (70) found a rapid increase in the dissolved P concentration of both surface runoff and tile drainage following grazing (figure 5-3). The study area was grazed in three sections by 100 dairy cattle at a stocking density of 121 cattle per acre per 24-hour period over 36 hours at the beginning of August, as part of the normal grazing plan for the farm. The tiles flowed continuously on this farm from April to September. The concentration of dissolved P in both surface runoff and tile drainage increased 1 day after grazing, with maximum values attained after only 1 week of grazing (figure 5-3).

The potential for P movement during grazing may be further enhanced by increased microbial activity mineralizing organic P adjacent to deposited manure (71). Also, organic P from manure may be more readily leached through a soil profile to shallow aquifers (9, 31, 69).

The assumption by many that pasture systems are more natural and thus, less likely to create environmental problems, may not be valid. Current evidence points to potential water quality problems if pasture systems are not managed properly. Many of the assumptions related to pastures and environmental impacts are based on experience with pasture systems in which supplemental nutrients are not imported onto the farm. There is evidence to suggest that there is a significant correlation between animal density and water quality impacts from nutrients (18, 45). Because of this, there is a need to evaluate the effect on water quality of feed nutrients imported to pasture systems on high-density livestock farms. Particular emphasis should be placed on determining the critical stocking densities and durations as a function of pasture management and water quality.

Table 5-1. The effect of grazing on runoff and erosion from a pasture watershed in Ohio.

Pasture management	Period (Years)	Runoff in.	Runoff %[a]	Erosion lb/ac/yr
Continuous grazing with hay feed	12 (1974–1986)	4.7	11	2015
Rotational grazing in summer	3 (1986–1989)	0.55	1	130
No grazing	5 (1989–1994)	0.24	1	8

[a] *Percent of annual precipitation as runoff.*

Source: Olness, A. E., S. J. Smith, E. D. Rhoades, and R. G. Menzel. 1975. Nutrient and sediment discharge from agricultural watersheds in Oklahoma. Journal of Environmental Quality *4: 331–336.*

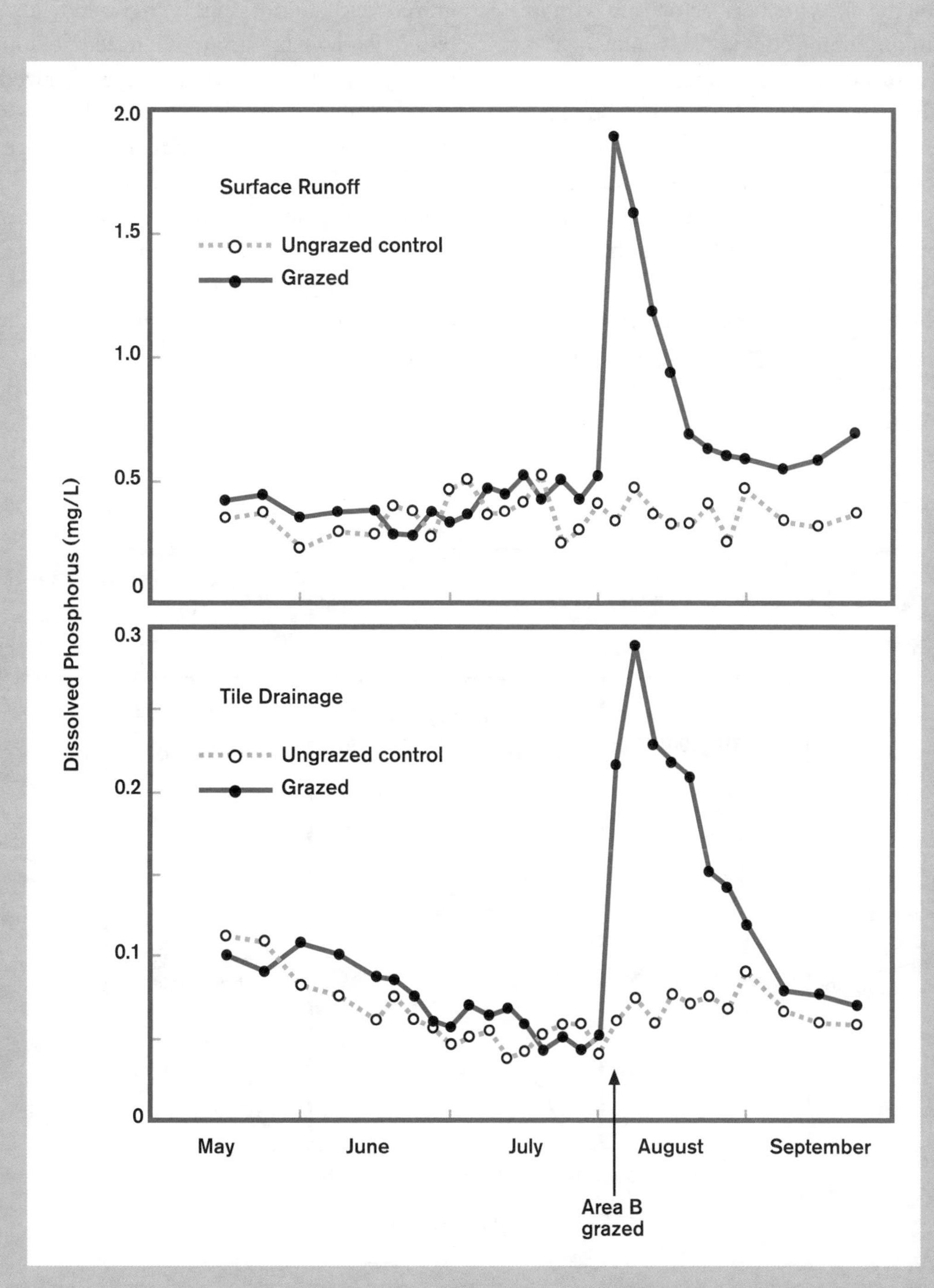

Figure 5-3. Mean dissolved P concentration in surface runoff and tile drainage before and after grazing on a dairy farm, Palmerston North, New Zealand.

Sources: Sharpley, A. N., and J. K. Syers. 1976. Phosphorus transport in surface run-off as influenced by fertiliser and grazing cattle. New Zealand Journal of Science *19: 277–282; Sharpley, A. N., and J. K. Syers. 1979. Loss of nitrogen and phosphorus in tile drainage as influenced by urea application and grazing animals.* New Zealand Journal of Agricultural Research *22: 127–131.*

Both hydrologic and chemical factors and level of nutrients imported in feed will likely influence these critical stocking factors. In terms of P transport, there will probably be a critical stocking level above which further poaching by animals will reduce infiltration sufficiently to dramatically increase runoff susceptibility and, thus, P loss. For example, one research team (70) found that grazing increased surface runoff (20%) and decreased tile discharge. Before grazing, the flow hydrograph for the discharge from two tiled fields was similar (figure 5-4a). In the first storm event after grazing, surface runoff was observed on the grazed field but not on the ungrazed field. As a result, tile discharge from the grazed field increased more gradually than that from the ungrazed field and peak flow from the grazed field was less (figure 5-4b). The hydrograph for discharge from the grazed field still increased more slowly than that from the ungrazed field 3 weeks after grazing (figure 5-4c). The reduction in tile drainage volume after grazing can be attributed to the disruption of macropores (large pores) at the top of the soil profile and to surface "pugging," which reduces infiltration capacity and preferential flow.

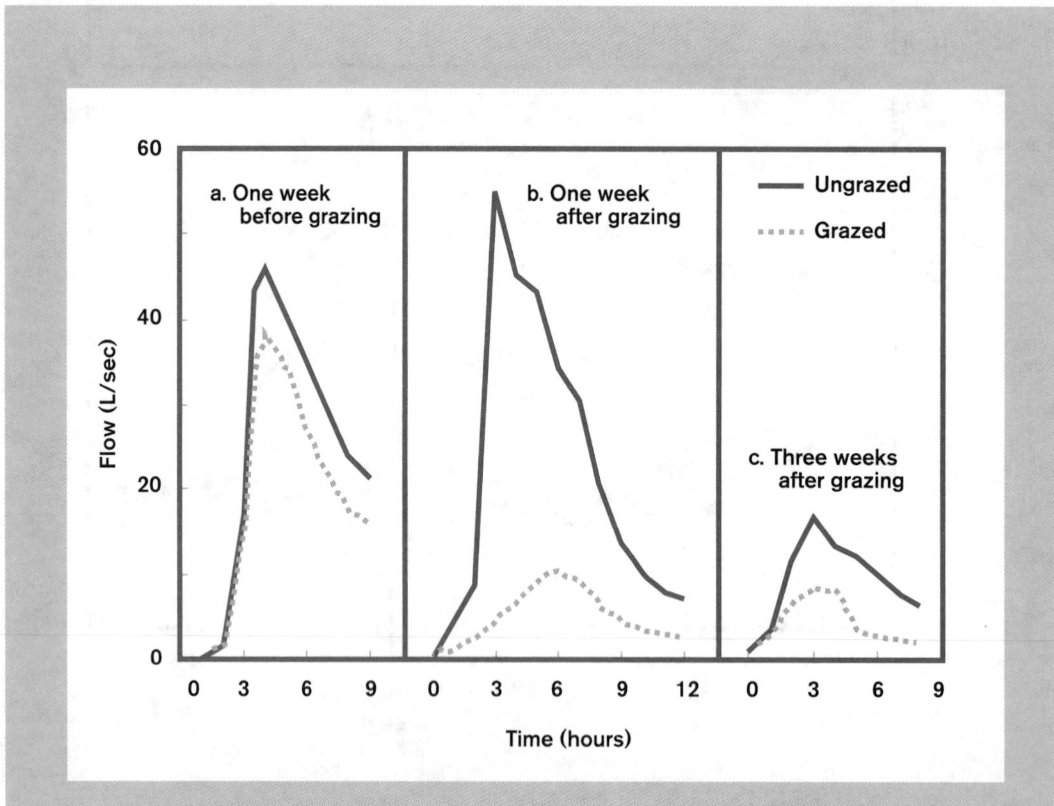

Figure 5-4. Tile discharge from ungrazed and grazed fields on a dairy farm, Palmerston North, New Zealand. (a) One week before grazing, (b) one week after grazing, and (c) three weeks after grazing.

Source: Sharpley, A. N., and J. K. Syers. 1979. Loss of nitrogen and phosphorus in tile drainage as influenced by urea application and grazing animals. New Zealand Journal of Agricultural Research *22: 127–131.*

PATHOGENS

Beef and dairy cattle are susceptible to a variety of bacterial and protozoan pathogens (organisms that cause disease) that can also make humans sick. For example, the protozoan parasites of *Giardia* species (spp.), *Cryptosporidium parvum,* and *Escherichia coli (E. coli)* cause scours in calves and diarrhea in humans (1). These pathogens are excreted in large numbers in the feces of symptomatic animals; infected animals without symptoms may excrete them at lower levels (22). Healthy cattle may also serve as carriers of human pathogens. For example, *E. coli* O157, *Campylobacter* spp., and *Listeria monocytogenes* have been documented in cattle feces (72, 88). These organisms, which cause bloody diarrhea, appear to be temporary components of the intestinal microflora of adult cattle. Consequently, animal health is not a reliable indicator of whether or not the animal is shedding these pathogens.

Limited, and often conflicting, information exists regarding the survival of specific pathogens in pasture environments. Laboratory studies indicate that fecal bacteria (*E. coli* O157 and *Campylobacter* spp.) die off rapidly in and on soils. However, watershed studies suggest that fecal bacteria may persist along river/stream banks and in channels. Increased concentrations of fecal bacteria are frequently observed in rivers/streams during storms because turbulent flow stirs up microbes in the water (43). *L. monocytogenes,* a bacterial pathogen, is always present in the environment (16). It can be isolated from soil, water, and decaying plant tissue, and from the feces of mammals, birds, fish, crustaceans, and insects. *L. monocytogenes* can survive and reproduce under most environmental conditions (39–111°F). Little is known about *Cryptosporidum* and/or *Giardia* survival in/on soils. Once in surface waters, however, these organisms can persist for several months (20).

A substantial body of literature exists describing the overland transport of microorganisms (particularly fecal bacteria) to surface waters (e.g., 6, 36, 58). The major factors controlling pathogen transport from pastures are excretion rates, livestock densities, proximity of animals to surface waters, microbial mortality rates, watershed hydrology, soil characteristics, vegetation type, and rainfall intensity and duration. Due to the variety and complexity of interactions, however, quantitative relationships have yet to be established regarding the extent or rates of pathogen transport.

SOIL CARBON DYNAMICS

The effect of grazing on soil organic carbon (SOC) has been extensively reviewed by Schnabel et al. (67). In this work they summarized the effects of various grazing management practices on SOC and animal production (table 5-2, p. 72). Schnabel et al. (67) found that while some specific practices such as using warm-season grasses and endophyte-free tall fescue have predictable effects on SOC, the effect on SOC of the overall practice of grazing management was unknown.

Recent unpublished research on cattle-grazing effects on SOC in the Southern Piedmont region of the United States has been summarized (41). During the first 3 years of steers grazing coastal bermudagrass, SOC increased at a rate of 1,338–1,606 pounds per acre per year. Under bermudagrass that was harvested as hay or left unharvested for conservation, SOC increased at a rate of only 267–357 pounds per acre per year. The similar, relatively low SOC accumulation rates under the hayed and unharvested management systems occurred because much of the aboveground, plant-derived C was not incorporated into the soil. In contrast, higher rates of SOC accumulation under grazing resulted from the return of much of the rumen-digested, plant-

Table 5-2. Implications of changes in grassland management on potential carbon sequestration and animal production.

Factor	Effect	
	C sequestration	Animal production
Replacing warm-season grasses with cool-season grasses	*Increase;* higher C:N; more SOC accumulation	Beef cattle–possible increased production, but increased management costs; Dairy cattle–decreased production from lactating animals
Replacing endophyte-infected tall fescue with noninfected tall fescue	*Decrease;* lower microbial degradation of C	Increased weight gains in beef cattle, also suitable for dairy production
Adopting intensive grazing	*Unknown;* increased plant biomass; increased SOC; increased forage quality	Increased weight gains and milk production
N fertilization–hayland	*Increase;* higher plant biomass production; higher SOC	Increased animal production per unit land area
N fertilization–intensively managed pasture	*Unknown;* increased biomass; increased SOC; increased forage quality	Increased animal production per unit land area
P fertilization	*Decrease;* more legumes in swards; lower C:N; less SOC accumulation	Increased animal production per unit land area

derived C to the soil; this material quickly became part of the SOC pool.

Following 15–19 years of cattle grazing Tifton 44 bermudagrass or Coastal bermudagrass in southcentral Texas, SOC to a depth of 7.9 inches averaged 32,736 pounds per acre (21). In a study of three sets of paired fields, one of which was hayed and one of which was grazed, the hayed fields contained 27,742 pounds SOC per acre. Most of the difference in SOC between grazed and hayed bermudagrass occurred in the surface 2 inches. Carbon in surface residue was also significantly greater under grazed (1,606 pounds per acre) than under hayed (1,070 pounds per acre) bermudagrass.

In another study, researchers (46) compiled a data set of 236 grassland studies worldwide that compared one or more attributes of grazed and ungrazed sites. Nearly all of the sites were rangeland that received less precipitation and fewer amendments and were far less productive than grasslands in the eastern United States. They concluded that grazing reduced aboveground net primary productivity, but total-system productivity was probably not affected by grazing due to compensation by the root system. Root mass for grazed vegetation was greater than for ungrazed in two-thirds of the studies, and belowground biomass production generally exceeds aboveground production. Only 37 of the 236 studies reported the effects of grazing on SOC. No relationship was found between SOC and whether or not grasslands were grazed, despite the lack of biomass removal from ungrazed areas in these studies (45).

In animal production systems, biomass is removed from grasslands either by grazing or by haying. A direct relationship exists between the level of SOC and annual additions of C to soil via crop residues (58). Consequently, SOC accretion should be greater under grazing than under haying because more C is returned to the soil. Grazing returns 60–95% of ingested nutrients as excreta to the pasture (81). In addition, stubble production with grazing can be up to 5% greater than with mowing due to animal herbivory stimulation of plant production that is not clearly defined (19, 85).

The effect of grazing versus haying on SOC in the Netherlands was mixed. Soil organic C in the surface 10 inches averaged 7,939 pounds per acre more under grazing than under haying at the end of 3 years of comparison (30). Little response was observed in SOC to fertilizer addition (223–624 pounds N per acre per year). In a later study at the same location, the effect of grazing and haying on SOC was small and inconsistent (29).

Grazing pressure and stocking method may influence SOC. Where moisture does not constrain yield, we would anticipate the following relative levels of SOC: management-intensive rotational grazing > intensive continuous grazing > extensive continuous grazing. During the first 3 years of cattle grazing on coastal bermudagrass, SOC increased significantly at a rate of 2.1 pounds per acre for each additional grazing day within the range of 243–486 grazing days per acre per year (79). More intensive grazing may enhance breakdown of aboveground litter and its incorporation into the soil (63). The increase in SOC may result from an increase in microbial populations as well as soil fauna and microflora contributions related to the microhabitat created by the addition of grazing animal byproducts. In contrast, where a moisture deficit limits production, as in western rangelands, long-term intensive grazing may damage the stand and cause a concomitant loss of SOC (17, 34). Fertilizer and lime additions and use of improved seed make intensively managed pastures more productive than extensively managed pastures.

Spatial redistribution of plant residue and excreta from foraging areas to camping areas can alter SOC. In a 15-year-old pasture in New Zealand, SOC in camping areas ranged from 37 to 84 ppm, but from 34 to 72 ppm in noncamping areas (50). In Georgia, SOC was also most concentrated near permanent shade and water sources (i.e., camping areas) in 7- to 15-year-old tall fescue pastures. Although some literature exists on redistribution of a few different nutrients in pastures (63, 90, 92), much more research is needed to understand SOC redistribution within pastures and its potential to sequester (bind and immobilize) C.

Replacing existing trees with improved grasses (74, 83) created most of the world's highly productive pastures. Consequently, a fair amount is known about the changes in nutrient cycling that occur when forestland is converted to pasture, or when unimproved pastures are managed to increase productivity. Major changes in organic matter content in surface soils often result from improved pasture management. Where organic C levels were initially low, large increases in organic C, N, P, and S have been measured for up to 50 years (64). Where organic C levels were high, the disruption resulting from conversion to pasture led to a reduction in organic matter (85). Organic matter levels rebounded during 25 years of pasture management. Increases have also been measured in soil microbial biomass, soil enzymes, and macroinvertebrates. One researcher (65) reported that the overall biological activity of the soil increases under pasture. Another research team (33) concluded that the greatest significance of increased levels of microbial biomass to pasture fertility is magnitude of nutrient fluxes through this large pool.

RIPARIAN AREAS IN GRAZING SYSTEMS

Riparian areas are lands adjacent to lakes, ponds, tidal areas, creeks, streams, and rivers where vegetation is strongly influenced by the presence of a permanent supply of subsurface water (8, 12). Pastures often extend into or surround riparian areas, and uncontrolled livestock grazing in riparian areas can increase sediment, nutrient, and pathogen load into the water body (83).

Much of the sediment comes from bank erosion along streams where riparian vegetation has been substantially lost (12, 92). Excessive sediment in the bed of a stream can destroy spawning grounds by smothering incubating fish eggs and newly hatched juveniles. It also destroys the habitat for many beneficial aquatic insects that are a food source for fish. Suspended sediment as low as 200 ppm can abrade fish gills, reducing their ability to absorb oxygen and weakening affected fish. At 20,000 ppm, sediment can clog gill filaments, preventing water circulation through the gills and suffocating the fish (87).

Livestock may have favorite watering places or stream crossings to get to other areas of the pasture (10). These natural drinking and fording areas may not always be very stable. High banks may often serve as windbreaks and rubbing areas. Gullies are often sought for protection against cold, stormy weather (83) or as sunning areas on cold, sunny days. These gullies may occur in the upper reaches of riparian areas or as lateral extensions and support riparian vegetation.

Nutrients, bacteria, and organics are components of animal waste and can contaminate surface waters within pastures or downstream from pastures. In streams, high coliform bacteria counts come from direct contamination of the water body, from erosion of banks or shores, and from runoff from upper watershed sources (8, 11). This is especially the case where livestock are allowed to camp in areas adjacent to water courses. Manure and urine spots are more frequent, the soil surface is compacted, and vegetation is thin or absent in camping areas. This combination of concentrated animal waste and ideal runoff conditions can lead to significant downstream loadings of nutrients and pathogens during storm runoff.

Pastures with streams running through them should have alternative sources of drinking water; this dramatically lessens animal time in the stream (47). Sizable associated bottomlands are best fenced off as separate grazing units from upland sites. These fences can be built inexpensively using high tensile wire. One or two electrified strands are often sufficient for cattle. This way, upland areas are better used, and riparian zones are not overused. Less manure and urine are deposited close to or directly in the stream. Livestock can be cycled through these bottomland pastures at times least injurious to the forage and riparian vegetation on the site.

Grassed riparian buffers at least 35 feet wide on either side of the stream are best to prevent overland sediment flow and reduce sediment-borne nutrient loading (43, 88). This distance keeps livestock from concentrating their time in the shade along the forest riparian buffer strip and creating a nutrient hot spot close to the stream. Grassed riparian buffers should be periodically harvested by grazing or cropping at 4-inch residual stubble height to remove accumulated nutrients and maintain stand density (11).

The type of vegetation in a riparian area can have an effect on stream temperature. In forested riparian areas, where shade extends over most of the channel during midday, water temperatures are cooler than in unforested riparian areas (42). This can be very important in maintaining cold-water fish populations such as trout. Shading and its effects on stream temperature are also important for some warm-water game species, such as smallmouth bass. High stream water temperatures foster the survival of nongame fish, such as suckers and carp, at the expense of game fish (56). One scientist (56) reported water temperature increases of 41–55°F from a wooded to an unshaded section of the same stream. The cooling effect of shade trees works well for streams up to 25 feet wide (13) where at least 80% canopy cover over the stream can be maintained (56). With wider streams, it is difficult to get the canopy cover required to shade the stream during midday even with mature, tall trees (40). This is particularly true of north-/south-flowing streams. On east-/west-flowing streams, shade trees on the north bank do little to shade the stream (42). However, even large aquatic life benefits from shade that occurs along the banks. Although water temperatures are not moderated, the shade provides an area of escape that is out of direct sunlight (42). Tall herbaceous vegetation, such as reed canarygrass and shrubs, can provide significant shading on streams less than 10 feet wide to support healthy fish populations (46, 55). Not only are summertime water temperatures moderated in smaller streams with shaded banks but winter water temperatures are as well. In winter, streams in wooded areas versus streams in cleared areas are significantly warmer by perhaps as much as 13°F (92).

The change in vegetation from a forested riparian area to an open grassy pasture has implications beyond stream warming. Streams in mature forested riparian areas have many debris dams created by fallen limbs and logs. These create fish cover and pools that are relatively free of sediment. These debris dams also hold detritus, such as fallen leaves, that are fed upon by aquatic invertebrates, such as caddis flies (Trichoptera), that in turn become forage for fish (87). The loss of this specific type of detritus and blockage reduces the productivity of stream fisheries (59).

The ultimate decision to create a forested buffer along a stream in a pastured floodplain must be based on a number of considerations. First is the type of fishery desired and its feasibility given the stream's potential. Second is the length of existing forested stream reaches above and below the pasture under consideration. If enough forested stream reaches are already present, a break in that cover from time to time may not be very detrimental. (Shrub plantings along stream banks could be an intermediate measure between treelessness and a forest.) The third consideration is the amount of groundwater flowing into the stream along its length. Groundwater tends to constantly chill the stream water. The fourth consideration is the stability of the stream itself. If it is actively eroding its bed and banks, watershed protection measures and/or in-stream revetment work may be necessary to control erosion. This may not be feasible depending on the nature and magnitude of the problem. Control of lateral bank movement is important, especially in minimalist riparian buffers (those that are narrow and mainly grass). If not controlled, fences can be undermined as stream oxbows move. The fifth consideration is the size of the pasture once the work is complete. If the buffer removes a significant portion of the pasture, a decision has to be made whether to continue pasturing it at all or whether just to create a grass buffer area. The sixth consideration is the amount and cost of fencing and stream crossings needed to create an effective vegetative buffer and useful pasture.

Forested or grassy riparian buffers do have some value for wildlife. However, the widths required for fullest wildlife benefit can be quite large, up to 900 feet or more. This would generally require a complete change in land use for most northeastern stream corridor pastures. The narrow grassy riparian buffers envisioned for aquatic habitat and sediment reduction are actually death traps for ground-nesting birds. Although nesting numbers are high, so is mortality due to the narrowness of the buffer areas and their attractiveness to predators that use them as protected travel lanes (46).

Wisconsin studies have demonstrated that continuously and rotationally stocked pastures actually supported more diverse bird communities than did grassy riparian buffers (46). On these pastures, livestock had access to the streams along their entire length through the pastures. Plant diversity was greater on the pastures than in the grassy buffers (55). Natural succession of plant communities tends to go quickly to tall, grazing-intolerant grass species if grass areas are left ungrazed (80), as they were in the Wisconsin study. However, invasion by undesirable plant communities and noxious weed control in riparian buffers can be an impediment to producers to establishing these buffers.

A riparian buffer acts principally by improving water quality through reduction of sediment and sediment-borne nutrients. A riparian buffer is much less efficient in removing soluble P from surface runoff or groundwater (13, 42). Some studies have found that neither grass nor forested riparian buffers were very effective in removing soluble P (13, 42). N removal is dependent in large measure (up to 80%) on denitrification. For this to occur, the soils in the riparian buffer must be poorly drained, highly organic, and anaerobic most of the time (13). N removal by the riparian buffer, forested or grass, may be as little as 4% of the total N exported from the contributing watershed if the buffer has a very narrow area that supports denitrification (42). Grassed buffers where denitrification could occur were effective in removing about 40–60% of various forms of N. This was accomplished in the first 25 feet of the buffer (42). Riparian areas subject to overflow may actually release accumulated nutrients sequestered in sediment and litter to overlying floodwaters when inundated (13).

Fencing to Exclude Livestock from Streams and Riparian Buffers

Livestock that defecate and urinate into stream and near-stream areas have the potential to contribute significant loads of nutrients over time. For instance, an average dairy cow can defecate up to 15 times and urinate 12 times daily, with a single defecation containing an average of 0.23 ounce N and 0.000635 ounce P (Sharpley and Stout, unpublished data, 90). Although urine does not contain significant concentrations of P (no pun intended), mean concentrations of N in urine range from 8 to 10 parts per thousand (ppt), with an approximate N load of 0.95 ounce in each Holstein urination (90).

To assess the possible benefits to water quality of excluding livestock from streams and riparian buffers, an observational study was conducted to quantify nutrient inputs from pastured cattle with stream access on a dairy farm in New York (36). Cattle were pastured 6 hours daily for approximately 180 days (May–October). Annually, approximately 4.2 pounds P (1.2 ounces per cow) were deposited directly into the stream in manure, while an additional 11.2 pounds P (3.2 ounces per cow) were deposited within 131 feet of the stream. Because near-stream soils are highly susceptible to surface runoff and flooding, nutrients applied to these areas are at high risk of enriching stream flow (36). On an areal basis, annual total P input from grazing cows defecating directly in the stream was 0.3 pound P per acre per year; within near-stream areas (within 131 feet of the stream channel) it was 1.1 pounds P per acre per year. These total P inputs are appreciably greater than the 0.12 pound per acre per year value reported for nonpoint sources in the 448-square-mile Cannonsville watershed, in which the study area is located (49, 67). This indicates that cattle in and near streams can be significant sources of nutrients in the water. The direct input to streams of P from defecating cows (i.e., 0.3 pound P per acre per year) is within the range of losses measured in runoff from grassed and cropped fields that received up to 44.6 pounds P per acre per year (i.e., 0.27–3.3 pounds P per acre per year) (68).

In addition to observing cattle behavior, a rapid biological assessment (using U.S. Environmental Protection Agency protocol (5)) was conducted to compare riparian and aquatic conditions within the pasture with those immediately adjacent to the pasture (36). Biological indices (i.e., species richness and abundance) calculated within the pastured reach of the stream were lower than those in areas outside the pasture.

Because flowing waterways intercept many acres of pastures across the Cannonsville watershed, exclusion of cattle from riparian buffers could reduce the direct loading of P and N to streams (36).

Even though numerous federal, state, county, and nongovernmental entities currently subsidize stream bank fencing, farmer participation in these programs is mixed. These programs may include stipulations concerning reimbursement, maintenance, and upkeep that cause farmers to balk. In addition, riparian exclusion may result in various secondary effects that are not subsidized. Farmers are often concerned about the loss of productive pasture land. Because riparian areas serve as watering sources to livestock as well as sanctuaries from oppressive heat, an alternative to these services must be provided. The layout of paddocks may require investments in infrastructure such as stream crossings and bridges. Once installed, stream bank fences must be properly maintained and replaced to ensure effective exclusion. This is a consideration after flooding of riparian buffers has occurred. General wear and tear on the fences by livestock must be continuously monitored.

Stabilized stream crossings and water accesses should be placed where pastures straddle streams and the riparian area is not fenced. Canadian work suggests that livestock will generally linger only to drink for 1–3 minutes at these sites, although on hot days cow may loiter in stream crossings for hours (10). When affordable, livestock water is provided using off-stream or off-pond watering facilities. This is especially true if pond or stream water is of questionable quality for livestock. In areas of actively eroding stream channels, livestock exclusion alone will not mitigate stream bank erosion (84).

Controlling Environmental Risks from Pastures

When improperly managed, pastured livestock can pose environmental risks similar to those caused by poorly managed confinement feeding operations. Several keys to good pasture management will reduce or prevent damage to environmental quality: proper plant nutrient and animal nutrition management, proper stocking rate and grazing pressure, and keeping livestock out of surface waters.

Leaching to groundwater and denitrification to air can result in major losses of N when the amount of N from fertilizer or supplemental feeds exceeds the forage stand's ability to use the imported N. Ammonia volatilization can be a significant loss of N to the air from urine spots, urea-based fertilizers, or spreading of manure that has not been composted.

These problems can be minimized by using legumes to fix N for grasses and by limiting the application rate of mineral fertilizer and manure N to that which the forage stand can effectively use. To minimize N excretion in urine, supplement livestock with a highly digestible carbohydrate source to maximize use of rumen-degradable protein and increase animal protein production so that it's in line with the protein supplied by the forage. Feed supplemental protein on pasture only up to the animals' requirement.

Provide supplemental feeds only in amounts that will allow the production of manure nutrients that the pasture can use and animal production will remove. Feeding high levels of supplemental feed produces similar results–nutrients imported to the pasture–as overfertilizing with N, P, or K. P as well as N can build up and cause surface water quality problems. Overfertilizing with K or K supplements in feed could ultimately lead to animal health issues.

Minimize soil compaction by maintaining a healthy sod using sod-forming species in the forage mix and by maintaining proper soil pH and fertility. Maintain soil cover for a healthy earthworm population and for other soil fauna and flora that will help maintain good soil structure. If needed during wet weather limit livestock access to pastures on soils subject to compaction. Management that increases soil carbon will help control compaction.

Keep animals out of surface water sources and use movable watering systems and riparian buffers. Flash graze grass buffers or use ungrazed forested buffers.

Reduce availability of shade from single trees, where livestock will gather. When heat is stressful to livestock use shade pastures that contain many trees at a uniform spacing, use movable shade panels, or keep livestock in the barn during the heat of day, allowing grazing when stressful heat has passed.

SUMMARY

Grazing is first and foremost an animal production system whose main purpose is to make the producer money by efficiently cycling nutrients and energy through ruminant animals to produce animal products. As with confinement production systems, grazing will have impacts on the environment and care must be taken to ensure that these impacts are acceptable to society. In contrast to confinement systems, however, grazing systems do not necessarily allow the producer to have the same degree of control over nutrient and energy flows. Consequently, if environmental impacts of grazing systems are to be minimized, constant care must be taken to employ the controls available to the grazing manager.

Some key management practices that can help control the potential negative environmental impacts of grazing are:

- Use of legumes as the pasture N source. This will reduce N leaching and help control stocking rate. Cattle eat more legume forage so fewer animals are needed to eat the forage there.
- Use of moderate energy supplementation to improve N utilization by the animal. This will reduce N excretion in urine and consequently reduce N leaching, provided that stocking rates are not increased excessively.
- Use of stream bank fencing and off-stream watering systems to reduce animal access to streams. This will reduce direct defecation and urination into streams and reduce stream bank degradation.
- Maintaining healthy pasture swards through proper stocking rate, grazing pressure, and soil fertility management. This will help prevent soil erosion and pasture degradation through compaction.

These and other management practices are discussed in detail in related chapters on soil fertility (chapter 3) and nutrient management (chapter 4).

Chapter 6
Invertebrate Pests, Weeds, and Diseases of Forage-Livestock Systems

Robert A. Byers, William S. Curran, Barbara W. Pennypacker, and Badruddin Ali Khan

Insects, weeds, and plant diseases are occasionally important in forage crops for both hay and pastures. Permanent pastures usually have few problems after establishment with plant pests, plant diseases, or weeds. Insects and other arthropod pests of grazing animals are economically important in permanent pastures and sometimes difficult to control.

Knowledge of the grassland invertebrate community (insects, mites, spiders, earthworms, millipedes, centipedes, slugs, etc.) is necessary to understand food webs and energy flow in pastures. Grasses and forbes are the basis of pastures and provide the nutritional requirements of the large grazing vertebrate animals such as cattle, sheep, horses, and goats, as well as the invertebrate grazers. Some invertebrates are exclusively herbivores (eat only plants), but others are omnivorous, feeding on living and decaying plants and dead insects, and sometimes preying on other insects. Other invertebrates (decomposers) are specialists in using decaying plant and animal material, and dung. A few invertebrate groups are exclusively predatory (spiders, centipedes, some rove beetles) or parasitic (certain wasps and flies) on other insects. Some invertebrates (earthworms, ants, and termites) are "ecosystem engineers" (13, 42), turning and drastically altering the physical structure of the soil. Some invertebrates are important in pastures because they affect the grazing animals directly. Biting flies, mites, and ticks infest or annoy grazing animals and cause losses in animal production. All of these invertebrates influence the energy flow in pastures and net primary production (plant photosynthesis – respiration and other factors reducing production) (62). Invertebrates, diseases, and weeds reduce net primary production by limiting growth or by competition for resources.

COMPOSITION OF THE INVERTEBRATE POPULATION

The invertebrate community of pastures is made up of a great number of species. The more species in a community, the richer or more diverse it is. If most species are equally abundant, then the population has a high degree of evenness. Dominance is most often the situation, with a few species being very abundant and other species occasionally or rarely present (53).

FACTORS REGULATING THE RICHNESS AND EVENNESS OF GRASSLAND INVERTEBRATES

Pasture plants have defenses against herbivores. Some plants have physical barriers or chemical defense mechanisms to herbivorous invertebrates. Physical barriers include simple and glandular hairs, waxy or tough leaves, or unattractive leaf colors. Chemical nutrients in plants, both quality and quantity, directly affect insect abundance and richness. Behavior-modifying plant chemicals such as attractants, repellents, and inhibitors interact with insect pests to help regulate populations (67).

Natural control of invertebrate herbivore populations is important in pastures. Many insects are susceptible to diseases caused by bacteria and fungi or are prey for parasites and predatory invertebrates (e.g., spiders, ground and rove beetles) and vertebrates (e.g., birds and mice). Competition between insect species and between individuals of the same species for food resources regulates both species richness and evenness. All of these factors interact to determine the size of invertebrate herbivore populations, which are usually at a level that leaves most of the net primary production for large herbivores (grazing animals). However, occasionally, the balance in these regulating factors is disrupted and invertebrate herbivores such as insects and slugs become pests in pastures.

LEGUME PESTS

A small metallic blue weevil, *Ischnopterapion virens,* (also known as white clover weevil or an apionid weevil) was recently introduced to the United States (41). First detected in 1994, it has spread to seven states in the Northeast (Pennsylvania, Maryland, Virginia, New Jersey, Delaware, New York, and Connecticut). This insect is widespread in Europe and feeds exclusively on all types of clover. Adult weevils have a body that is pear-shaped and about ⅛ inch long, and have a long snout. They make pinholes in the leaflets in both spring and fall. Damage from adult feeding is not severe, but larvae feeding internally in stems and stolons can destroy up to 80% of infested stems. Infested white clover plants are chewed to bits; infested red clover plants lose stems and leaves, reducing the size of the plant. It has not been determined if this insect will limit white clover production in pastures in the United States, but there is evidence from Europe that it may be an important pest. The insect is locally abundant, is expanding its range in the region, and eventually will occur in every pasture in the humid eastern region.

The clover root curculio feeds on roots of clovers and alfalfa and has been a pest for more than 100 years in the United States. A native of the Mediterranean region, this introduced weevil has spread across the entire country and parts of Canada (12). It occurs wherever clovers and alfalfa are grown. Adult weevils are brown and ¼ inch long; they feed by making notches on the edges of leaves. Damage by adults is not usually a limiting factor on growth, but larvae infest the nitrogen (N)-fixing nodules of roots, reducing the N fixation capability of the legume. Older larvae prune small roots and gouge the surface of taproots. Plant pathogens, such as the fungus-causing fusarium wilt, enter the wounds caused by larvae and rot the taproots (43). The combination of larval feeding and fungal disease infection leads to a reduction in growth and premature death of the plants. The legume plant density is reduced through time and either weeds invade where legumes once flourished or an increase in the use of mineral fertilizers is required to maintain production. Close grazing of pasture reduces the egg-laying habit of the clover root curculio (20) and may be a control method to reduce the root injury.

The potato leafhopper is a very small (⅛ inch), light green jumping insect that does not survive the winter in the northern United States. Each spring, it migrates north from the Gulf Coast area and infests more than 100 different plant species during the growing season. This leafhopper feeds in pastures by piercing the stems of clover and alfalfa and secreting a sheath of protein material, which blocks the phloem tubes. This blockage causes sugars to increase in the leaves, which interferes with photosynthesis and causes leaves to die, turning yellow in the case of alfalfa and red in the case of clovers. Some scientists (25) believe that "leafhopper burn" is a saliva-induced response. Heavily infested plants during summer are stunted and yield less forage. Many of these infested plants are so stressed that they die the following winter. Chemical sprays at the appropriate time before yellowing and stunting occur will control the insect. Some glandular-haired alfalfa varieties have been released with partial resistance to this pest. There are no resistant clover varieties.

GRASS PESTS

Several groups of insects and other invertebrates are occasionally pests of grasses. Field crickets, ground crickets, and gray garden and marsh slugs are important pests during establishment, especially with no-tillage introduction of legumes into established sods.

Many species of leafhoppers other than the potato leafhopper increase in abundance during summer on most grasses. They stress plant growth by removing plant nutrients at a time when most cool-season grasses are already under water and temperature stress. The impact of leafhopper feeding has not been well studied (14).

Soil-dwelling white grubs, such as the Japanese beetle and others, billbugs, and wireworms (larvae of click beetles) are major pests of turf grasses (72) but only occasionally infest pastures. Their potential to cause losses in grass production is not well documented.

PESTS OF TURF GRASSES ARE NOT AS IMPORTANT IN PERMANENT PASTURES

The current theory is that the extensive plant diversity in most pastures in the Northeast allows for greater invertebrate diversity or richness (2). More niches occur in these diverse pastures for predators and parasites of plant and animal pests. This greater diversity among the plant population permits or tolerates some grazing by invertebrate herbivores, but not all plants are affected and there is enough biomass for large animals. One scientist (21) estimated that about 10% of biomass is lost annually to invertebrate herbivores. In less diverse plant systems there could be problems. Where monocultures of perennial ryegrass have been used for increased production in the United Kingdom, white grubs and leatherjackets can be major problems. However, another scientist (2) stated that perhaps a single ecological theory will not account for the variety of responses of such a broad group of animals as the insects.

MANAGING GRASSLANDS TO CONTROL GRASSLAND INVERTEBRATES

Grazing management has a profound effect on growth rate and species composition of pasture plants, which in turn influence invertebrate herbivores and decomposers. Some scientists (17) have found that some grasshopper species are more abundant in heavily grazed pastures, whereas others increase in low- to medium-grazed pastures. These scientists thought it possible to modify grasshopper abundance through regulation of grazing intensity. Grazing management also affects populations of natural enemies indirectly, by its influence on the abundance of nectar- and pollen-producing plants. Nectar and/or pollen serve as food for adult parasitic wasps and flies, which attack invertebrate herbivores. Trampling by large animals can limit soil invertebrates such as slugs and has been used as a control method in New Zealand (26).

Mowing pastures affects pests and beneficials by eliminating flowers, seedheads, etc., and by opening the plant canopy to sunlight, which increases soil temperature, ultimately affecting soil invertebrates. The amount of manure available for pests and beneficials is strongly influenced by stocking density.

GRASSLAND RENOVATION

Some invertebrates such as crickets and slugs can be devastating pests during no-till establishment of legumes in grass sods (15, 35) because the sod provides habitat for the pest and the legume seedlings are highly palatable. Potato leafhoppers can be a problem to seedling legumes in conventional tillage methods (16). Timing of planting is important in avoiding problems in pasture renovation. Early spring planting is preferred to reduce the effect of slugs and potato leafhoppers during establishment (15).

LIVESTOCK PESTS IMPORTANT IN PASTURES

Our ability to use improved fertilizers, irrigation, and varieties of grasses has helped in maximizing the number of cattle per unit area. Intensive cattle grazing has resulted in high concentrations of cowpats per unit area, which are the primary source of dung-breeding insect pests of cattle.

Horn flies and face flies are two introduced pest species in pastures that breed in cattle dung. The blood-feeding activity of the horn fly worries cattle and results in poor feed conversion. The production losses due to horn fly have been estimated at more than $730 million annually (24). Face flies are non-blood-feeding pests. They land on the face of animals and feed on saliva, tears, and nasal secretions, causing animals to congregate and discontinue their normal grazing. Face flies can spread pinkeye bacteria between animals, and they also act as vectors of eye worm (a nematode). The estimated annual loss due to face flies in the United States exceeds $53 million (23).

Cattle dung is a natural habitat for a large number of pasture insects—as many as 457 species in North American pastures (6). Some are beneficial and act as natural enemies of horn flies and face flies. These regulate the pest fly populations directly as predators and parasites, and indirectly as competitors for a common food source (38). The most important beneficial insect group, the dung beetles, fragment, distribute, and bury cattle manure in pastures and therefore eliminate food for pest fly populations. Dung beetles have great economic value because of their dual roles in natural recycling of nutrients in pastures and in acting as a natural control agent of pests. It is estimated that total annual benefit of the dung-burying activity of these insects is more than $2 billion in the United States (27).

CATTLE DUNG IN PASTURES–A VALUABLE RESOURCE

Cattle dung is a valuable nutrient resource and should be incorporated quickly into the soil before the loss of N to the atmosphere occurs (27). In Africa, an entire cowpat is buried in a day by large numbers of dung-inhabiting fauna (76). In the United States, cattle produce more than one billion cowpats a day, and about 60–80% of them drop on pasture (19). Although dung beetles bury some cowpats, most remain on the surface undisturbed, where they dry quickly and become unattractive to insects. This is an indication that there are insufficient numbers of beneficial dung-feeding insects in pastures. Undisturbed cowpats also prevent the growth of pastures beneath them for the season. Rank growth at the outer perimeter of the cowpat is unpalatable, and cattle avoid eating it (76). A single cow may deposit 11 cowpats a day covering an estimated square yard of pasture area with manure (19). In about a 6-month grazing period, the excretions of a small herd of 50 cows could decrease an area of pasture by 5 acres, an area that could be used to produce one calf. If all these cowpats were buried in the soil within 48 hours, it would reduce the insect pest problem, add N and other nutrients to the soil, reduce nonpoint source pollution in the pasture, and increase the water retention ability of the soil underneath the cow pat. The dung-burying activity of insects can also increase the permeability of soil to water infiltration by 47–346%, depending on the soil type (64).

Role of Dung Beetles

The three distinct groups of dung beetles beneficial to the pasture ecosystem are the "paracoprids" (dung feeders), "endocoprids" (dung-burying beetles), and "telecoprids" (dung rollers) (7). The paracoprid beetles live inside the cowpat, where they raise their brood. This activity helps in the disintegration of cowpats and at the same time reduces the habitat of pest flies. This group of beetles makes the pat porous and accelerates dung decomposition on pasture.

The endocoprids dig tunnels directly underneath cowpats, bringing soil to the surface. They pack the tunnels they create with compressed manure and deposit their eggs. These manure packs, either in the form of pellets or dung balls, are essentially the brood balls. A single beetle egg is laid in each ball, which serves as food for the larva. Endocoprids add organic fertilizer to the soil, remove manure from the pasture surface, and bury eggs of pest flies. Quick burial of the dung in soil by these beetles helps conserve the N in the feces (28, 34).

Telecoprids are not abundant in North America, although they are prevalent in the tropical areas of the world, especially in Africa. The African dung roller beetles can make an entire cow pat disappear in 48 hours. They shape manure into the form of balls and roll them to the base of grasses away from the cowpat or deposit them in shallow pits.

Among the three groups of dung beetles, dung-burying paracoprids were successfully exploited by the Australians to control cattle dung in their pastures, and for controlling buffalo flies and bush flies (8). The success of the program prompted the introduction of these beetles into Hawaii for the control of horn flies, and eventually into the mainland United States (34). A total of 29 species of dung beetles has been introduced in the United States. However, only nine were successfully established (28).

Rove beetles are often found associated with dung and successfully breed in cowpats. Some species feed on fly eggs and can cause severe mortality of horn flies (45). One type of rove beetle is a parasite of face fly pupae. This beetle was introduced from Europe in 1966 (23), and it has recently become established in Pennsylvania.

IMPACT OF PESTICIDES ON BENEFICIAL PASTURE INSECTS

The metabolites and residues of systemic insecticides and broad-spectrum anthelmintic veterinary drugs (used to control parasitic worms) used for the control of external cattle pests and internal parasites are excreted into the manure and can affect beneficial dung-feeding beetles and retard dung degradation on pasture (75). Slow-release bolus formulations used during late spring and summer may have a greater impact on the immature stages of the dung-inhabiting nontarget insects than injectable formulations. The bolus formulations release insecticide residues in the manure for up to 140 days, whereas the injectable formulations have a shorter period of activity (30, 31). It is important that producers choose products that will have minimum impact on beneficial pasture insects. Pesticides usually are not selective and often kill more than the target pest species. Several researchers have shown that cattle dung treated with insecticides for the control of dung-breeding pest flies was slow to decompose due to lack of sufficient beetle activity in the treated dung. A report from the Worldwide Fund for Nature warns that the routine use of pesticides is undermining the vital agricultural and ecological role of the beneficial insects.

The economic value of dung beetles to the pasture ecosystem cannot be ignored. There is a renewed interest in employing these insects in sustainable animal agricultural practices. In Australia, producers are redistributing imported dung-burying beetles once released in the 1970s.

PLANT DISEASE MANAGEMENT IN PASTURES

Plant disease is a concern in monocultures and simple mixtures managed for hay production, but is of less concern in pastures. The plant diversity that characterizes pastures in the northeastern United States reduces the spread of plant pathogens by reducing the proximity of susceptible plants. Species diversity is very effective in limiting disease epidemics (49). Grazing also reduces disease problems in pastures. Grazing animals remove diseased foliage from the pasture and thus limit secondary spread of airborne pathogens (48). Unfortunately, grazing does not limit root and crown rot pathogens. Indeed, overgrazing may exacerbate root and crown rot diseases (48) by reducing the plant's photosynthetic ability and thus the amount of carbon available for defense responses.

Disease Development

Management of plant disease can be expedited if the factors that cause disease development are kept in mind. Disease occurs only when a pathogen, a susceptible plant, and a conducive environment coincide (49). The absence of any of these factors will prevent the development of disease. Most of the fungal pathogens of forage plants require cool, moist conditions for growth and production of spores. These pathogens are primarily a concern during the spring and fall. If the pathogen is present and the weather conditions favor pathogen growth, disease will develop unless the plant has genetic resistance to that particular plant pathogen (see sidebar). In general, drought and heat inhibit disease development in forage plants.

Plant Disease Resistance

Genetic resistance is the primary means of controlling disease in pastures and hay fields. Plant resistance can be broadly categorized as either qualitative or quantitative. Qualitative resistance is controlled by a dominant gene and usually prevents symptom development. Unfortunately, qualitative resistance appears to put a great deal of selection pressure on the pathogen population for those individuals that can grow on the plant. This selection pressure results in the development of more virulent races of the pathogen that can defeat the qualitative gene and cause disease. Once the resistance gene is defeated, it is no longer effective and plant breeders must deploy a new resistance gene.

The other broad category of disease resistance is quantitative resistance. This is controlled by a number of small genes with additive effects. Quantitative resistance does not prevent the pathogen from attacking the plant, but does minimize the detrimental effects of pathogen presence (18, 57).

Disease resistance in heterogeneous crops such as alfalfa has a somewhat different meaning than in a homogeneous crop such as corn. A cultivar of alfalfa that is resistant to verticillium wilt, for example, would have at least 50% resistant plants. The remaining plants would have varying degrees of susceptibility to the pathogen. Loss of the susceptible plants would not have a detrimental effect on the hay crop because the remaining resistant plants would compensate by increasing in size.

Plant disease epidemics develop when the pathogen spreads to adjacent susceptible plants. When susceptible plants are abundant, the number of infected plants increases logarithmically, creating a threat to the yield potential of the field. Such a scenario is unlikely in a pasture because species diversity reduces the potential for a concentration of susceptible plants. In essence, species diversity buffers pastures against the development of economically significant levels of disease. If species diversity is lacking, then resistant cultivars of the desired forage species should be used when renovating or establishing pastures. Economics preclude the use of fungicides to control plant disease. This underscores the importance of resistant cultivars in disease management.

Plant Diseases and Pasture Establishment

Several plant diseases should be considered when establishing a new pasture. The fungi causing these diseases are particularly problematic on seedling legumes. Sclerotinia crown and stem rot is a threat to fall-seeded alfalfa (48). The fungus infects seedlings in the fall and remains active over the winter, often killing the infected plants. A poor stand the following spring is often attributed to winterkill, when in fact the culprit is *Sclerotinia*. Often, small black sclerotia can be found at the base of dead plants, confirming *Sclerotinia* as the pathogen. This pathogen is not a problem in spring-seeded alfalfa and red clover because the seedlings have age-related resistance by the time the fungus produces spores in the fall. However, spring establishment of alfalfa following a soybean crop that had a significant amount of white mold can be risky (see sidebar) (58). If a soybean/alfalfa rotation is favored, the alfalfa should be planted by mid-April to avoid infection by white mold (58).

Soybean/Alfalfa Rotation

Barbara W. Pennypacker and Marvin H. Hall

Soybean is susceptible to Sclerotinia stem rot, or white mold, as it is commonly called. The disease is caused by the soil-borne fungus *Sclerotinia sclerotiorum*. The pathogen produces large, black structures called sclerotia by which it survives adverse conditions. The sclerotia, which resemble rodent scat, can remain viable for up to 7 years in the soil. A field experiment was conducted to determine whether *S. sclerotiorum* from soybean would be a problem during spring establishment of alfalfa. The treatments were planting date (mid-April and mid-May) and pathogen. Sclerotia of the pathogen were scattered on selected plots, and other plots were protected by monthly application of a fungicide. The fungus infected alfalfa but did not cause significant yield reduction during the seedling year. The major effect of the pathogen appeared in the following spring. Overwintering was reduced in plots that were planted in mid-May and not protected with the fungicide (table A).

Table A. Effect of *Sclerotinia sclerotiorum* from soybean on spring establishment of alfalfa. Data represent percent winter survival, evaluated in March.

	Planting Date	
Pathogen	mid-April	mid-May
No *Sclerotinia*	79.8 a[a]	71.2 a
Sclerotinia	79.6 a	49.8 b

[a] *Values within a column with different letters are statistically significantly different at P = 0.02.*

The results indicate that timely planting is important if alfalfa follows a soybean crop that was severely infected with *S. sclerotiorum* (58).

In addition to Sclerotinia rot, the diseases to keep in mind when establishing a pasture are Pythium and Phytophthora root rots (18, 49, 71). These fungi cause root rot and collapse of seedling legumes. Phytophthora can also cause root rot on mature plants. Pythium and Phytophthora root rots are mainly a problem in areas where the soil lacks good drainage and is too wet (49). It is important to establish drainage in poorly drained fields prior to planting. To protect seedlings against these two pathogens, seed should be treated with an approved fungicide (50). Phytophthora root rot affects mature plants; therefore, select alfalfa cultivars with resistance to this pathogen when establishing a pasture (71). When selecting any grass or legume forage variety, ensure that it is resistant to plant diseases commonly occurring in the area.

Plant Diseases in Established Pastures

Several diseases may become problematic as pastures age. In particular, a number of root and crown rots affect legumes and limit the stand life of red clover, alfalfa, birdsfoot trefoil, and white clover. These complexes are not well understood and a number of organisms have been associated with them. Currently, the best way to manage these diseases is to encourage good growth by keeping pH and fertility levels in the pasture near optimum (18, 49, 71). In addition, grazing should be managed so as to allow the plants sufficient time to replenish root reserves of carbohydrates. Such a strategy will permit the legumes to express inherent resistance they may have to these microorganisms.

Alfalfa diseases such as anthracnose stem and crown rot and the vascular wilts, namely Fusarium wilt, Verticillium wilt, and bacterial wilts, are best controlled by planting resistant cultivars (18, 71). Although these diseases are ubiquitous in the Northeast, cultivars with good levels of resistance are available. Viruses are present in clovers in the Northeast, but their effect on yield and plant longevity is not known. Leaf spot symptoms caused by a number of fungi are frequently present on legumes and grasses. These diseases are primarily managed through grazing or mowing, which removes the diseased leaves from the pasture and reduces the chance of subsequent infection (50). As with the root rot complexes, it is important to encourage good plant growth by maintaining adequate fertility and pH.

Tools for Disease Management

Species diversity combined with resistant cultivars when available should minimize most disease problems in pastures. Maintaining pasture fertility and the correct pH will encourage good plant growth and provide the resources the plant requires for the expression of resistance to various pathogens. Grazing management is important due to the detrimental effect that overgrazing can have on plant performance. Overgrazing stresses the plant by removing most of the photosynthetic leaf area and, in the case of orchardgrass, the basal stem tissue that stores carbohydrates (33). Carbohydrate reserves are needed for regrowth and also for the expression of disease resistance. Well regulated grazing has the advantage of removing diseased leaves and stems from the pasture without hampering the plant's ability to withstand both biotic and abiotic stress. Agronomic management that contributes to vigorous plants and species diversity coupled with the use of resistant cultivars and sound grazing management will limit disease problems in pastures.

WEED MANAGEMENT IN PASTURES

Weeds are troublesome in many ways. They reduce crop yield by competing for water, light, space, and soil nutrients. Weeds can replace desirable grass species, filling in gaps or voids and reducing yield and overall quality of pasture and forages. They can produce allelopathic substances that are toxic to crop plants. In addition, certain plants such as poison hemlock, white snakeroot, or black locust have toxic properties that can cause livestock injury or loss under certain circumstances. To plan an effective weed management program, a producer must be able to identify weeds present and understand how weed biology and ecology affect where weeds are found and their value or detriment.

Competition

Weed competition in pasture systems has not been extensively examined. In addition, the bulk of the competition research in higher rainfall areas such as the Northeast has been conducted in Australia or New Zealand, not the United States. In pasture, weed control decisions are largely based on visual thresholds and intuition. There is rarely reliable biological information or cost-benefit analysis to support weed management decisions. Without question, weeds can directly compete with forage grass or pasture to reduce its nutritional value and longevity. However, the impact of weed species, density, and soil and climatic factors is not well established in pasture systems. In general, biennial and perennial weeds are probably the biggest weed problems for pasture producers. Both produce seed each year, potentially starting new infestations. In addition, perennial weeds such as ironweed, Canada thistle, and multiflora rose reproduce from underground roots or rhizomes. Perennial rooting structures can survive for several years in the soil and are often unaffected by occasional mowing or livestock grazing. Pasture-invading weed species should be assessed for competitive ability (potential to reduce desirable forage species); invasiveness (potential to multiply and increase); yield, quality, and value relative to desirable forage species; and cost and effectiveness of control (cultural, mechanical, and chemical).

Key Points about Weed Competition in Forages

- Weeds that emerge with the crop in the spring are generally more destructive.
- Control problem weeds for the first 60 days after seedling establishment.
- Weeds that emerge beyond 60 days after establishment will not influence that year's forage yield.
- Later-emerging weeds may still influence forage quality.
- Winter annual weed competition in early spring is most damaging to early-season forage yield.
- Broadleaf weeds that are biennial or perennial are generally more competitive than grassy weeds.
- For all weeds, assess competitive ability, invasiveness, nutritive value, and potential to control.

Weed Quality

Unlike most grain or fiber crops from which weeds are separated at harvest, weeds are often harvested along with forage crops, potentially reducing quality, or in the case of pasture, they remain in the field, where they continue to interfere with desirable forage. Reductions in quality of forage in weed-infested pasture are often in the form of lower protein content, feed digestibility, or even reduced intake by the animals. Although weeds do have some feed value, this value differs among species. The feed value of weeds has not been extensively studied for many pasture species. However, based on traditional forage quality measures—crude protein and digestibility—many weeds are nutritious and readily digested by livestock during the growing season (9, 10) (table 6-1). Wild carrot, a common pasture weed in some fields, has about 16% crude protein in the vegetative stage (74). Common yarrow has only about 10% crude protein during the flowering stage (74). Grassy weed quality can be similar to that of grass forage. In general, weedy grasses have about 75% of the quality of forage legumes. The main problem with weedy grasses is early maturity. Weeds with woody stems or flower stalks such as tall ironweed, yellow rocket, and curly dock have protein levels that are about half to two-thirds of a legume forage and are slightly less in quality than forage grasses. Like grass and legume forage species, the quality of weeds is better during their vegetative stages and decreases as the plants flower and mature (table 6-1). Finally, even though some weeds are highly nutritious and digestible, ruminants may avoid grazing these plants because of taste, smell, or toxicity (47, 63).

Table 6-1. Percent crude protein and in vitro dry-matter digestibility (IVDMD) for selected broadleaf and grassy weeds and three forage species. Range of values corresponds to samples evaluated from the vegetative stage to fruiting stage (broadleaves or forbs) or heading (grasses). Palatability for these weed species was not determined.

	% Crude protein	% IVDMD
Broadleaf weeds		
Henbit	20.1–16.2	78–75
Pepperweed, Virginia	31.9–17.1	86–63
Dock, curly	29.9–16.1	73–51
Pigweed, redroot	23.9–10.6	73–64
Jimsonweed	25.1–16.5	72–59
Grassy weeds		
Cheat	23.4–13.8	81–61
Barley, little	23.6–13.8	82–62
Panicum, fall	19–7.2	72–54
Foxtail, yellow	17.5–14.3	73–57
Crabgrass, large	14.3–6.4	79–63
Forage species		
Clover, ladino ("Regal")	27.2–23.2	83–81
Fescue tall ("Kentucky 31")	22.1–12.5	78–67
Rye ("Wrens Abruzzi")	27.9–13.4	79–70

Sources: Adapted from Bosworth, S. C., C. S. Hoveland, and G. A. Buchanan. 1985. Forage quality of selected cool season weed species. Weed Science *34: 150–154; Bosworth, S. C., C. S. Hoveland, G. A. Buchanan, and W. A. Anthony. 1980. Forage quality of selected warm season weed species.* Agronomy Journal *72: 1050–1054.*

Poisonous Plants

Many plants contain potentially poisonous substances that may be toxic to livestock if consumed. In addition, certain plants may be problematic because they irritate the stomach when eaten, cause photosensitization, and/or have disagreeable tastes or odors in meat, milk, or milk products (37). If you suspect livestock poisoning, call a veterinarian immediately. If death occurs, the stomach contents should be examined for consumed herbage. Identify the suspected plants and remove livestock from the grazing area until all poisonous plants have been removed or destroyed. Table 6-2 lists some common weeds of the Northeast and their poisonous properties.

Key Points about Weed Forage Quality and Poisonous Plants

- Some weeds have excellent nutritive quality.
- Weeds in the vegetative stage of development usually are more nutritious than more mature weeds.
- Regardless of weed quality, livestock may avoid grazing certain plants because of taste, smell, or toxicity.
- Some plants contain potentially poisonous substances that may be toxic to livestock if consumed. Properly identify potential problem weeds and consult with a veterinarian if necessary.

Problem Weeds

Weeds are grouped into three categories (annuals, biennials, and perennials) based on their life cycles. Annuals complete their life cycle within 1 year and reproduce only by seed. They may produce as few as 100 seeds per plant or as many as 500,000 seeds per plant, depending on species and growing conditions. Annual weeds are classified as winter or summer annuals. Winter annuals germinate in the fall, overwinter as a rosette or small clumps of leaves, and complete their reproductive cycle in the spring or early summer. These weeds are more likely to be found in perennial forages and pastures where soils are not disturbed over the winter. Some examples of winter annuals are given in table 6-3 (p. 92). Summer annuals germinate in the spring and set seed in late summer or fall. Summer annuals thrive when summer annual crops like corn or soybean are grown. They can also be a problem during the establishment year for new spring forage seedings or if established forages become thin or irregular. They complete their lifecycle in the late summer or fall. Some examples of summer annual weeds are provided in table 6-3.

Biennial weeds live during two growing seasons and reproduce only by seed. The first year consists of vegetative growth in which the plants produce a rosette or loose clump of leaves and a fleshy taproot. The second year involves both vegetative and reproductive growth, whereby an elongated flower stalk is produced. Because these weeds require 2 years to complete their life cycles, they are found in areas of low soil disturbance such as waterways, pastures, hay crops, and fencerows. Biennials are rarely a problem in annually cultivated soil because plowing usually destroys them. However, established biennials often survive field cultivation or disking and may continue to be a problem in reduced or no-tillage production. Examples of several biennials are provided in table 6-3 (p. 92).

Perennial plants live for more than 2 years and generally reproduce by vegetative structures as well as by seed. Vegetative reproduction occurs through structures such as rhizomes, tubers, bulbs, or budding roots. However, not all perennials reproduce vegetatively. Simple perennials reproduce only by seed production and emerge from the same vegetative structure every year. Dandelion is an example of a simple perennial

Table 6-2. Selected poisonous plants of the Northeast.

Common name	Poisonous plant part	Problem/ symptoms	Toxic ingredient	Toxicity
Bouncing bet	Leaves and stem	Delayed for several days; depression, vomiting, abdominal pain, diarrhea	Saponin	Amount equivalent to 3% (dry wt.) of sheep wt. killed within 4 hr.
Buttercups	Leaves and stem, especially in flower; dried hay loses toxicity	Anorexia, weakness, convulsions, breathing difficulty, death	Protoanemonin	Toxicity reported to vary with species, age, and habitat.
Cherry, black	Leaves (wilted leaves are worse), stems, bark, and fruit	Anxiety, staggering, breathing difficulty, dilated pupils, bloat, death	Cyanogenic glycosides	Less than 0.25 lb leaves (fresh wt.) can be toxic to 100-lb animal.
Clover species	Vegetation	Hairballs; Sweet clover: nose bleeding, anemia, abdominal swelling	Coumarin with sweet clover	Varies
Fern, bracken	Entire plant	Dullness, fever, bleeding, loss of appetite, and salivation	Glycoside thiaminase	Cattle fed 50% bracken for 30–80 days had toxic effects.
Garlic, wild	All plant parts	Tainted milk and meat	Only toxic in large quantities	–
Hemlock, poison	All plant parts	Salivation, vomiting, diarrhea, weakness, paralysis, trembling, dilation of pupils, convulsions, and coma	Coniine and others	0.5–4% (fresh wt.) equivalent of cattle wt. is toxic.
Jimsonweed	Entire plant (seeds are most toxic)	Thirst, mood swings, convulsions, coma, death	Solanaceous alkaloids	0.06–0.09% (dry wt.) equivalent of animal body wt. is toxic.
Locust, black	Leaves (especially wilted), seeds, and inner bark	Weakness, depression, anorexia, vomiting, and diarrhea	Phytotoxin robin, glycoside robitinum	Bark extract and powder in amount equivalent to 0.04–0.1% of animal wt. is toxic to horses. Cattle are 10 times more tolerant.
Milkweeds	Stems, leaves, and roots	Muscle tremors, spasms, bloat, difficult breathing	Glycosides and galitoxin	Varies
Mustards	All parts (especially seeds)	Oral and gastrointestinal irritation, shaking, salivation, abdominal pain, vomiting, and diarrhea	Thiocyanates, irritant oils, and nitrates (large quantities generally necessary for toxicity)	–
Nightshade species	Vegetation, unripe fruit	Loss of appetite, salivation, weakness, trembling, paralysis	Solanine	Toxic at 42 ppm (LD_{50}).
Pigweed species	Foliage	Kidney disease, weakness, edema, rapid respiration	Nitrates, nitrate oxalates, unknown	Sheep, hogs, and young calves most susceptible.
Pokeweed, common	Entire plant, especially roots	Gastrointestinal cramps, weakened pulse, respiration, salivation	Saponins, phytolaccine, and phytolaccotoxin	10 or more berries can result in toxicity to humans.
Snakeroot, white	Leaves and stem	Constipation, loss of appetite, salivation, rapid respiration; toxin passes through milk (milk sickness)	Trophine alkaloid	Varies from 1 to 20% of animal body wt. Toxin cumulative.
St. Johnswort	Flowers and leaves	Photosensitivity	Hypercin	Uncertain

Sources: Adapted from Gardner, F. P., R. B. Pearce, and R. L. Mitchell. 1985. Physiology of Crop Plants. *Iowa State University Press, Ames. pp. 290–291; Hardin, J. W. 1973.* Stock-Poisoning Plants of North Carolina. *Bulletin No. 414. Agricultural Experiment Station, North Carolina State University, Raleigh; Hill, R. J., and D. Folland. 1986.* Poisonous Plants of Pennsylvania. *Pennsylvania Department of Agriculture, Harrisburg, PA.*

Table 6-3. Selected examples of annual, biennial, and perennial weeds that infest forage crops.

Annuals		*Biennials*	*Perennials*		
Winter annuals	**Summer annuals**		**Simple perennials**	**Creeping perennials**	**Woody perennials**
Bluegrass, annual	Black medic	Burdock, common	Chicory	Ironweed, tall	Autumn olive
Brome, downy	Crabgrass, large	Carrot, wild	Dandelion, common	Garlic/onion, wild	Brambles
Chickweed, common	Fleabane species	Cockle, white	Dock species	Groundcherry, smooth	Honeysuckle, tartarian
Dead nettle, red	Foxtail species	Hemlock, poison	Pokeweed, common	Heal-all	Rose, multiflora
Henbit	Jimsonweed	Knapweed, spotted		Hemp dogbane	
Horseweed/ mare's tail	Lambsquarters, common	Mullein, common		Horsenettle	
Lettuce, prickly	Nightshade, eastern black	Parsnip, wild		Johnsongrass	
Pennycress, field	Panicum, fall	Thistle, bull		Milkweed, common	
Pepperweed, field	Pigweed species	Thistle, musk/plumeless		Nettle, stinging	
Radish, wild	Ragweed, common			Nutsedge, yellow	
Shepherd's-purse	Smartweed, Pennsylvania			Quackgrass	
Yellow rocket	Velvetleaf			Red sorrel	
				Thistle, Canada	
				Wirestem muhly	
				Yarrow, common	
				Yellow toadflax	

(table 6-3). Creeping perennials often reproduce by both vegetative structures and seed production. Canada thistle is an example of a creeping perennial (table 6-3). Finally, woody perennials may be either simple or creeping, but they also produce secondary growth or a woody structure that enables them to become very large and usually aggressive. Multiflora rose is an example of a woody perennial that is also creeping (table 6-3).

Although perennial weeds are most prevalent in areas of reduced soil disturbance, some are well adapted to row crops. Pasture and hay production systems are often ideal environments for the growth and spread of perennial weeds. Managing perennial weeds is generally more difficult because of their multiple reproductive systems. Consider both vegetative structures and seed when dealing with perennials.

Weed Management

Managing weeds in pasture systems should begin long before crop establishment. Certain types of weeds are potentially serious problems for forages, so it is important to eliminate them in advance. In particular, perennial broadleaf weeds such as dandelion, curly dock, Canada thistle, and quackgrass are much easier to manage prior to planting a forage crop. In addition, biennial weeds, including musk thistle and burdock, should be eliminated before establishing forage. If these weeds are not removed before the seeding is done, they can persist for many years. The cost of controlling weeds before or at the time of seeding should be considered an investment that will be returned for the life of the forage.

Cultural Weed Management

Cultural practices that aid in the control of weeds include anything that makes the crop more competitive against weeds. In the establishment year, these include proper seedbed preparation, optimum planting date, proper fertilization, planting at higher densities, using the correct seeding rate, choosing high-quality crop seed that is free of weeds, and selecting adapted species and varieties for the region. In general perennial grasses are more competitive with weeds than legumes.

Providing a seedbed at planting that is free of live weeds can be accomplished either with tillage or the use of an herbicide. It is important that emerging forage species are not subject to competition for limited resources as they try to gain a foothold in the early weeks of establishment. In addition, emerged vegetation can harbor certain insects or pathogens that could attack young susceptible forage seedlings.

Date of planting can influence the species and numbers of weeds. Most grass and legume forage species are relatively slow to establish. Consideration of planting in spring versus fall should be based on weed history and anticipation of potential problems. For example, if you're establishing pasture in a field that was previously planted in corn or some other summer annual crop, then summer annual weeds will likely be the biggest weed threat during establishment. Late summer may be a better time for establishment in this situation. In spring seedings, early planting before summer annual weeds emerge gives the new forage seedlings every advantage. With late summer seedings, planting before September, the month that winter annual weeds generally begin to emerge, is better than later plantings. The weed species present in a field along with its potential severity may help determine the best time for planting.

In established pasture systems, prevention is the most important tool for managing weeds. Research shows that pasture weeds are controlled by increasing forage competition (59). In fact, crop growth rate stands as the single best measure of plant response to weed competition in forages (65). Maintaining dense competitive forage is a key to preventing weed invasion and interference. Weeds are opportunistic. Germination and establishment are favored by open areas and by disturbance. Weeds can be suppressed by overseeding with desirable forage species when necessary to keep open areas at a minimum. Rotational grazing minimizes traffic effects on pastures, and avoidance of overgrazing ensures that forages remain competitive with weeds. Testing soils for nutrients and annually fertilizing pastures keep forage stands healthy and competitive. Controlling harmful insects or pathogens is essential because they weaken forage stands and allow weeds the opportunity for establishment. Priority in monitoring pastures to locate weed infestations should be on controlling small infestations so they do not expand.

Preventing weed infestations also means preventing dispersal of seeds or vegetative structures into uninfested areas. Vehicles, humans, wind, water, birds, and livestock can spread weed seeds. Animals may disperse seeds by picking them up in their coat, fur, or between the pads of their feet. Cattle are known to readily pick up burrs of several weeds when grazing forested range (22). Infested animals should be cleaned regularly, particularly new animals that may be carrying new weed problems. Ruminants ingest weed seeds in the field, and somewhere between 5 and 15% pass safely through sheep, goats, cattle, and deer (56). Producers should be cautious of feed or hay infested with noxious weed seed. In the western United States, certified weed-seed-free forage is required on public lands by federal land agencies (55).

Key Points about Cultural Weed Management

- Consider seedbed preparation, planting date, fertilization, planting population, and high-quality crop seed. Select adapted species and varieties.
- In established pasture systems, prevention is the most important tool for managing weeds.
- Overseed with desirable forage species when necessary to keep open areas at a minimum.
- Rotationally graze to minimize traffic effects and do not overgraze.
- Soil test for nutrients and fertilize to keep forage stands healthy and competitive.
- Prevent dispersal of seeds or vegetative structures into uninfested areas.

Mowing and Hand Removal

Once forages are up and established, systematic mowing will help control weeds. Repeated mowing reduces competitive ability, depletes carbohydrate reserves in the roots, and prevents seed production. Some weeds, mowed when they are young, are consumed and found palatable by livestock. Mowing can kill or suppress annual and biennial weeds. Mowing can also suppress perennials and help restrict their spread. Mowing at a height above the grass seedlings when weeds are 8–10 inches in height can reduce shading by weeds. A single mowing will not satisfactorily control most weeds. However, mowing three or four times per year over several years can greatly reduce and occasionally eliminate certain weeds, including Canada thistle (66, 77). Mowing along fences and borders helps prevent the introduction of new weed seeds.

Hand removal may be the easiest, most economical method of control of some weeds when only a few weed plants are present. Removing the seedhead prior to seed dispersal is as effective as digging the entire plant for annual and biennial species. For perennials, it may be difficult to effectively remove all vegetative structures. Properly dispose of weeds after removal to prevent seed or vegetative structure dispersal. This may mean burning, burying, or composting the material.

Key Points about Mowing and Hand Removal

- Repeated mowing reduces competitive ability, depletes root carbohydrates, and prevents seed production.
- Mow at a height above the grass seedlings when weeds are 8–10 inches in height to reduce shading.
- If you see a new weed, dig it, pull it, or remove the seedhead prior to seed dispersal.

Herbicides

Herbicides provide a convenient, economical, and effective way to help manage weeds. They allow fields to be planted with less tillage, and allow earlier planting dates. Where herbicides are not an option, mechanical and cultural control methods become that much more important. In pasture systems, a number of herbicides are available for broadleaf weed control in grass forages. However, few herbicides are available for mixed grass-legume combinations or for the control of grassy weeds in grass forages. Before establishment, herbicide choices are limited to controlling emerged vegetation. Preplant soil residual herbicides are not common for pasture systems. Most herbicides for pasture systems should be applied postemergence to the weeds and crop once the forage is well established. In pasture systems, spot spraying may be the most economical alternative for scattered infestations of weeds.

Key Points about Using Herbicides in Forages

- Thin or irregular stands do not always thicken once weeds are removed. Be sure there are sufficient desirable species to fill in the gaps or overseed if necessary.
- Weeds tolerant of the herbicide may invade the space left by susceptible species, ultimately creating a more severe weed problem.
- If 50% or more of the stand is weeds, it is time to renovate or rotate to a different crop.
- If weeds become a problem in established forages, several herbicide options are available. Many products have harvesting, feeding, or grazing restrictions following their use.

Young annual weeds in the seedling stage are most susceptible to control with herbicides. Spraying biennial weeds is most effective in the rosette stage prior to bolting. Perennials are most susceptible to control with systemic herbicides in the bud to bloom stage or in early fall. Most herbicides for broadleaf control in grass stands should *not* be applied to seedling grass until visible tillers are present. Established forage grasses and legumes are more tolerant than seedlings to herbicides. Most herbicides have haying or grazing restrictions following application. For specific herbicide recommendations, consult current cooperative extension weed management guides or manufacturer product labels.

Biological Control

Biological control can be defined as the deliberate introduction or manipulation of a pest's natural enemies, with the goal of suppressing the pest population (79). It has been used to manage insects, vertebrates (mice and rats), pathogens, and weeds. The goal of biological control is not to eradicate the target weed, but rather to exert enough pressure on the pest to reduce its dominance to a more acceptable level (79). Biological control should be cost-effective, environmentally safe, self-perpetuating, and well suited to an integrated weed management program. However, the limitations of biological control are: 1) it is a long-term undertaking, 2) its effects are neither immediate nor always adequate, 3) only certain weeds are potential candidates, and 4) the rate of failure for past biological control efforts has been fairly high.

Biological control tools for weeds have included insects, mites, nematodes, pathogens, and grazing animals (e.g., sheep and goats). Insects and mites have historically been the most important biological control tools for weeds (44). The emphasis for developing biological control agents for weed management has been on western rangeland and natural areas. In the Northeast, several weeds, including bull and musk thistle, Canada thistle, purple loosestrife, mile-a-minute, and garlic mustard, are receiving attention because of their invasive nature. In addition to the several promising insect biocontrol tools outlined in table 6-4, rust fungi are being evaluated for the management of several weeds, including the knapweeds and the thistles. Biological weed control may have a greater impact on managing problem weeds in pastures systems in the future.

Key Points about Biological Control and the Use of Grazing Animals

- Biological control tools for weeds include insects, mites, nematodes, pathogens, and grazing animals.
- Biological control can be cost-effective, environmentally safe, self-perpetuating, and well suited to an integrated weed management program.
- Biological control is a long-term undertaking, and it has certain limitations. It is not immediate nor always adequate, only certain weeds are potential candidates, and the rate of failure can be high.
- Grazing will not in most cases eradicate a mature infestation of weeds.
- Combining mowing or an herbicide application with grazing can provide a wider window for control. When using herbicides on pasture it is essential to follow the herbicide's label restrictions for grazing.
- Biological weed control may have a major impact on managing problem weeds in pasture systems in the future.

Weeds and Grazing Animals

Grazing management can be used to minimize the spread of certain weeds and to control large infestations. However, in most cases grazing will not eradicate a mature infestation of weeds. For grazing animals to be useful for weed control, the area must be fenced to allow adjustments to grazing pressure. The ability to concentrate stock on weed infestations at some stages of growth or times of the year, and the ability to keep them off pasture or weeds at other times is often the key to weed control (60). Cattle, sheep, and goats are the most common animals used to graze pasture. However, horses may also be used. Pigs sometimes graze grass and legumes, but their weed control activities are more associated with their rooting behavior. Domestic birds will also eat grass and have been known to graze weeds selectively (60). Combining ruminant grazing (e.g., cattle, sheep, and goats) with other weed management tools can offer an integrated approach that may be very cost-effective.

Cattle

Grazing that is restricted to one class of stock such as cattle leads to particular weed problems because some weedy plants are less palatable to some classes of stock. At low grazing pressures cattle prefer grasses and legumes and tend to avoid forbs and shrubs; hence constant grazing by cattle may reduce grass forage and promote forbs and shrubs, some of which may be weeds. For some weeds, cattle can provide effective control partly because of their grazing patterns and partly because their hooves can damage young, tender, and emerging shoots. In general, selective grazing and overgrazing by cattle usually creates more problems, such as bare patches in pastures, which allow invasion of new weed seedlings. Rotational stocking with properly controlled grazing pressure is one means of reducing this problem.

Sheep

Grazing by sheep is a major method of biological weed control on dryland farms in Victoria, Australia, where they are used during fallow periods and to reduce weed seed production before cropping (1). Sheep prefer broadleaved plants (forbs) to grasses and shrubs. In Saskatchewan, continuous summer-long sheep grazing reduced the number of leafy spurge seeds from 3,500 to 15 seeds/1.2 square yards after 8 years (11). In this experiment, sheep grazing had no effect on leafy spurge stem density

Table 6-4. Potential biocontrol organisms for selected weeds of the Northeast.

Weeds	Biocontrol organism	Remarks and Citation
Knapweed, spotted	Root gall beetle	Larvae feed on root, weakening plants (61)
	Seedhead weevil	Larvae feed on seeds in seedhead (69)
	Seedhead fly	Larvae feed on seedhead (70, 78)
	Seedhead gall fly	Larvae feed on seedhead (70)
Loosestrife, purple	European beetles	Larvae feed on young buds, leaf, and leaf tissue (4)
	Weevil	Adults feed on leaves and larvae damage roots (5)
Rose, multiflora	Rose rosette disease	Mite-vectored virus (73)
	Rose seed chalcid	Wasp adults lay eggs in seeds, rendering them sterile (73)
	Rose stem girdler	Larvae girdle and kill canes (73)
Thistles-bull, Canada Thistles-musk, plumeless	Seedhead weevil	Feed on the developing seedhead; relatively effective where established; less effective on plumeless and Canada thistle (46, 68)
Toadflax, yellow	Weevil	Adults feed on buds, flowers, and seed capsules (36)
	Beetle	Adults feed on young shoots and flower buds; larvae feed inside seedhead (54)

for the first 3 years, after which densities declined dramatically. Today, sheep are being used to control leafy spurge along several major rivers in Montana (55). In addition to spurge, sheep have been used successfully to control Canada thistle (39). Control by grazing alone requires intensive grazing of the young, soft, aerial thistle shoots in spring, which is not usually possible because of pasture feed surpluses during that time. However, combining mowing or an herbicide application (keeping in mind herbicide label grazing restrictions) with grazing can provide a wider window for control.

Goats

Goats are capable of controlling a large number of spiny and prickly weed species totally untouched by sheep and cattle (3). In a North Carolina study, 12 goats/acre alone or 7 goats/acre mixed with cattle mostly eliminated multiflora rose and some other weeds from an abandoned orchard after four grazing seasons (51). In the same experiment, desirable forage species increased in number over time. Goats have also been used successfully for general brush control on abandoned farmland in Vermont (80). Goats are known to eagerly consume flowering thistle plants, but are not attracted to the vegetative rosette (60).

Combining small ruminant grazing with other weed management tools has considerable promise for certain weed species. For example, grazing Canada thistle with sheep and goats during the spring and fall followed by a fall application of an appropriate herbicide could have a greater impact on the weed than either tactic alone. In addition, adding sheep or goats to a cattle enterprise for control of weeds or to help clear land of undesirable vegetation can potentially be profitable. In a West Virginia study, 3-year variable costs for brush clearing with goats were estimated at about \$13.50/acre versus \$54/acre for mechanical cutting and \$240/acre for herbicide (52).

Integration

An integrated program that combines cultural, mechanical, chemical, and perhaps biological control tools can provide effective economic weed management in pasture systems. Different tactics can be combined if it is important to consider how weed lifecycle and other growth characteristics affect management options. Prevention is the most important consideration for managing weeds in established pasture systems. Some general guidelines for managing annual, biennial, and perennial weeds are provided in table 6-5.

RECOMMENDATIONS AND SUMMARY

Preparation of Site for Pasture Establishment

Managing pests in pastures begins long before pasture establishment. Perennial broadleaf weeds such as horse nettle and Canada thistle may be potentially serious problems and should be eliminated in advance of planting. It is important to establish drainage in poorly drained fields or to select disease-resistant forages to avoid losses from Pythium and Phytophthora root rots in establishing pasture in spring.

Table 6-5. Management guidelines for some problem weeds of pasture systems.

Weed	Cultural/mechanical	Chemical	Other
Annuals			
Winter annuals (Mustard species, common chickweed, etc.)	Mow after bolting to prevent seed production. Maintain a dense sod to suppress seedling emergence.	Apply an effective herbicide in fall or spring prior to bolting.	Most winter annuals emerge by late fall. A smaller percentage will emerge in early spring. Prevent seed production to prevent spread.
Summer annuals (Pigweed species, common lambsquarters, common ragweed, etc.)	Keep pasture full and competitive. Mow after bolting to prevent seed production.	Apply an effective herbicide in early summer.	Prevent seed production to prevent spread.
Biennials			
Biennials (common burdock, bull and musk thistle, poison hemlock, etc.)	Mow after plants have bolted but before seed is set to prevent seed production. Remove or dig out individual plants by hand.	Apply an effective herbicide to rosettes in the spring or fall. Several insect biocontrol tools may help with thistles in the future.	Prevent seed production to prevent spread.
Perennials			
Creeping perennials (Canada thistle, horsenettle, etc.)	Mow to suppress vegetative growth and prevent seed production. Co-grazing with sheep or goats will control some species.	Spray with an effective systemic herbicide at bud to bloom stage or in early fall prior to frost.	Most perennials spread by both seed and vegetative structures. Biocontrol tools may help in the future.
Woody perennials (multiflora rose, autumn olive, etc.)	Mow to suppress and prevent seed production. Remove roots by hand or with heavy equipment. Co-grazing with sheep or goats will control some species.	Spray with an effective systemic herbicide at bud to bloom stage or in early fall.	Rose rosette disease and insect biocontrols could help manage multiflora rose in the future.

Pasture Establishment

Prevention is the most important tool for managing establishment problems. Insects, weeds, and plant diseases are occasionally important in the establishment of forages for both hay crops and pastures. Knowledge of the grassland invertebrate community is necessary to understand food webs and energy flow in pastures. Recognition of symptoms of pest damage can be essential to successful establishment of pastures. Several plant diseases must be considered during establishment; for example, Sclerotinia crown and stem rot is a threat to fall-seeded legumes. To protect seedlings from Pythium and Phytophthora root rots, seed should be treated with an appropriate fungicide. Early spring planting avoids many problems with invertebrates, such as slugs and potato leafhopper, because the crop can be established before these pests become active and cause injury to seedlings. Using a complex mixture of plant species, including disease-resistant species or varieties, in pasture establishment also limits problems with invertebrates, diseases, and weeds. Plant species diversity buffers pastures against the development of economically significant levels of disease and invertebrate pests. To plan an effective weed management program, producers must be able to identify weeds present and understand weed biology. Weeds that emerge in early spring are generally more destructive and should be controlled in the first 60 days after planting. Winter annual weed competition in early spring is the most damaging to yields, and broadleaf biennial and perennial weeds are the most competitive. Cultural practices that make the crop more competitive against weeds and more tolerant of disease and invertebrate feeding include proper seedbed preparation, optimum planting date, proper fertilization, using an adequate seeding rate, and choosing weed-free crop seed.

Maintaining Established Pastures

Diseases, weeds, and invertebrates may become problems as pastures age. The best way to manage diseases such as crown and root rots is to keep fertility and pH levels near optimum. Planting resistant varieties best controls some diseases such as anthracnose stem and crown rot and the vascular wilts. Unfortunately, there has been less progress with resistant varieties to control invertebrate pests such as clover root curculio and potato leafhopper. These pests are best managed with cultural practices such as grazing and mowing at proper intervals to maintain vigorous stands. In fact crop growth rate is also the best single measure to prevent weed invasion and competition. Rotational grazing is preferred to keep forages competitive and to keep open, bare soil areas to a minimum. However, intensive grazing management results in severe concentrations of cowpats in pastures that are breeding sites for animal pests. Avoiding overgrazing when at all possible helps control animal and plant pests, diseases, and weeds. Mowing and hand removal of problem weeds can suppress weeds, keep forages competitive, and eliminate insect and disease problems.

The Future for Grazing Management and Pasture Pests

Maintaining plant diversity, exercising moderation in grazing pressure, and control of large animal pests and parasites are the most important strategies of the future. An integrated program that combines cultural, mechanical, chemical, and biological control tools can provide effective economic weed management in pasture systems. Species diversity combined with disease-resistant species and cultivars should minimize most disease problems in pastures. Invertebrate species composition should be monitored to detect any changes in population richness and evenness brought on by management strategy changes or invasion of pastures by exotic foreign invertebrates. The challenge will be to sustain pasture productivity without upsetting the balance between herbivorous species and their natural enemies, beneficial decomposers of dung and plant material, invasive weeds, and detrimental plant diseases. Management of all these factors will avoid pest outbreaks.

CHAPTER 7
Establishing Forage Stands

Lester R. Vough, Marvin H. Hall, and Edward B. Rayburn

Productive, high-quality pastures result from the application of sound pasture management practices, which are based on knowledge of the available soil resource, how plants grow, the soil and climatic requirements of plants, and how plants respond to grazing. Grazing managers and advisors need to understand the adaptation, growth, and production characteristics of the forages to be grown just as much as they need to understand breeding, nutrition, and management of animals.

Obtaining thick, vigorous new forage stands depends on selecting high-quality seed, timing of seeding, proper seeding techniques with equipment precisely adjusted for seeding rate and depth, seedbed preparation, and environmental conditions. Other key factors include proper soil pH and fertility, and adequate control of weeds and insects.

Invested in every new forage stand is not only the cost of seed but also the costs for time, labor, and capital in seedbed preparation and planting. In addition, feed shortages that may occur when a new seeding is lost must be made up in some other way at added cost. Also, a planned rotation may be disrupted by the failure of a forage stand, and the soil could be left unprotected and susceptible to erosion. All these factors add up to a considerable investment. Thus it behooves producers to follow proper seeding and establishment procedures to encourage vigorous forage stands. A number of management practices are necessary for successful establishment.

PLANNING FOR NEW SEEDINGS

The need for and approach to new seedings will be determined by the goals and objectives for the pasture program, the type of animal enterprise and its forage quality requirements, soil type and characteristics (drainage, topography, fertility, etc.), composition of the existing vegetation, and equipment available for seeding and maintenance practices. Limiting factors must be identified early in the planning process and certainly before seed is purchased.

Preparations for seeding must begin as much as 2 years prior to the actual planting of the seed, especially for no-till seedings in which lime and fertilizer cannot be incorporated and mixed into the soil. Most old, permanent pastures and existing grasslands needing reseeding will require lime, fertilizer, and weed control. These treatments should be applied 6–24 months prior to seeding.

When initiating a pasture improvement or renovation program, begin with the most productive soils available. The better the soil, the greater the potential improvement from reseeding or renovation. Do not attempt to reseed or renovate the entire pasture acreage at one time. Animals must be removed from seeded areas until the plants become adequately established to withstand grazing. If the costs of lime and fertilizer recommended by a soil test are prohibitive, it is far better to reduce the area to be renovated and apply the full amount needed than to spread inadequate amounts over a larger area. Each succeeding year, renovate additional areas as required and as finances allow. It may be 3–4 years until renovation of the entire acreage is accomplished.

If converting cropland to pasture, residual herbicides from previous crops must also be considered. Herbicides used on crops such as corn and soybeans may have restrictions for planting forage legumes and grasses following these crops. The restriction periods can be up to 2 years or more in some cases (new herbicides can be up to 30 months). For example, problems of triazine carryover may be encountered with forage seedings following corn, particularly in fields following 2 or more years of no-till corn (8). It may be necessary to alter herbicide rates in the final year of a corn-soybean rotation, to switch to shorter residual herbicides, or to use Roundup Ready or Liberty Link crops to reduce herbicide residuals. Contact the local extension office or commercial field representative for specific information and recommendations.

In a rotation of 2 or more years of no-till corn followed by small grain and then no-till forage, lime should be applied to correct any pH problems prior to the last year of corn. This will release any triazine herbicides bound in the soil while the field is still in corn (11). Applying the lime immediately prior to seeding the forage crop can result in triazine toxicity to forage grasses and legumes and stand failure. Where possible, some type of tillage prior to planting the last year of corn or the small grain helps to move lime and fertilizer into the plow layer and to reduce potential herbicide injury.

CONTROLLING EXISTING PERENNIAL BROADLEAF WEEDS

Perennial broadleaf weeds are usually present in older pastures, especially those that have been poorly managed, and may be present on cropland to be seeded to hay or pasture. Successful forage establishment of mixed grass and legume stands depends on elimination of these weeds prior to the time of seeding, through tillage, herbicide application, or both. For best results the weed control program should begin 6 months to 1 year before seeding. Treatment will kill most desirable forage legumes that might be present in the existing pasture, but this step is necessary to kill undesirable broadleaf weeds.

The type of herbicide to be used will depend on the composition of the existing vegetation and the kind of seeding. If a satisfactory stand of grass exists and the main purpose of the seeding is to introduce legume species, then selective herbicides such as dicamba and/or 2,4-D should be used for broadleaf weed control (see chapter 6). Early fall (2–4 weeks prior to the average killing frost date) is the most effective time period for application of herbicides to deep-rooted, difficult-to-kill perennials such as dandelions, curly dock, milkweed, dogbane, and Canada thistle. If desirable forage species are not present and all of the existing vegetation needs to be killed, then a nonselective herbicide such as glyphosate should be used.

LIMING AND FERTILIZING

Persistent, high-yielding forage stands are always associated with a favorable pH and high fertility. The importance of soil testing and following a carefully designed fertility program is emphasized in chapter 3.

For conventionally tilled seedbed seedings, soil samples should be taken to the depth of the plow layer. For no-till seedings, two sets of samples should be taken—one from the 0–2-inch depth to determine surface pH and fertility and the other to the normal plow depth. The 0–2-inch sample is a must for fields that have been in no-till corn production and will not be plowed following the last year of corn. Surface applications of nitrogen (N) fertilizers frequently cause this layer to become quite acidic.

Ideally, lime should be applied 6–12 months prior to seeding; it should be thoroughly incorporated into the plow layer to neutralize soil acidity. With no-till seedings, surface applications should be made 1–2 years ahead of seeding to allow for movement into the soil profile. A study (10) found that yields of legumes and total forage in the seeding year were greater with incorporated than surface-applied lime; however, in succeeding years, annual yields and 6-year total yields were not significantly different due to liming method.

Phosphorus (P) level is especially critical during establishment. It is also commonly a limiting factor on unproductive, poorly managed pastures. A readily available supply of P within reach of the roots of young seedlings is essential for normal root development and seedling establishment. On soils of low to medium P content, drills capable of banding 15–30 pounds/acre of P_2O_5 fertilizer directly below the seed at seeding promote establishment (13, 14). Broadcast application of P is not as effective as banded application. It takes at least four times as much broadcast P to give the same growth response as banded application does when soil test P is low.

The demand for potassium by young seedlings is relatively low. It is much more important once stands are established; high levels are essential for maintaining productive, long-lived stands.

Application of fertilizer N at seeding has been shown to increase yields when soils are low in N (less than 15 parts per million [ppm] soil nitrate-N) or when organic matter is less than 1.5% (9). When soil nitrate levels were greater than 15 ppm and conditions were favorable for effective nodulation (soil pH 6.2–7.5 and high populations of appropriate *Rhizobium* bacteria present), using N fertilizer in legume establishment did not result in economical yield increases. Rates in excess of 30 pounds N/acre generally have detrimental effects on rhizobial infection and N fixation by legumes in the stand (20, 21). N should never be applied in renovation overseedings because it quickly stimulates growth of the existing vegetation, leading to excessive shading of young seedlings and increased competition for water, light, and soil nutrients.

GRAZING OR CLIPPING CLOSE

The one time overgrazing is beneficial is when preparing a pasture for a complete renovation and reseeding or overseeding of weak pasture. Intensive, close grazing or frequent close clipping over the fall and winter prior to a late winter seeding will assist in suppressing competition from existing vegetation. Forcing livestock to closely graze unpalatable and/or low-quality pastures can reduce animal performance. This is unacceptable for growing animals or lactating dairy cows but is generally acceptable for mature, nonlactating livestock.

MATCHING PLANTS TO SOIL AND SITE CHARACTERISTICS

Many factors must be considered when selecting suitable grass and legume species. A wide range of grass and legume seed is available. Each forage species has its own particular characteristics that make it more or less suitable for a particular site and management purpose, as described in chapter 1. Many forage plantings fail or perform poorly simply because the species chosen for planting is not adapted to the site.

First and foremost when selecting species is the necessity of matching grasses and legumes to the characteristics of the soil on which they are to be grown. Soil type, drainage, moisture-holding capacity, fertility, pH, and winter hardiness all have an effect on plant species adaptation. Producers, farm supply personnel, farm advisors, and consultants often select or recommend species based on personal or industry preferences and biases without considering soil and site characteristics.

Here's an illustration of a rather common situation. A state extension specialist was asked to visit a newly constructed horse farm to investigate a pasture establishment failure. It was a beautiful waterfront property fully constructed with new barns, board fences, laneways, and roads—pretty much everything a horse owner would want—except green pastures. A typical "horse pasture" mix of about a half-dozen grasses and three or four legumes had been seeded, but by midsummer of the year after seeding, very few of the desirable grasses and legumes seeded remained. Pastures were mostly either bare soil or crabgrass and weeds.

It was readily evident that most of the soil on the farm was very poorly drained and unsuited for orchardgrass, bluegrass, and white clover, as the owners desired. Other species in the mixture such as perennial ryegrass and timothy were not adapted to the region's hot, dry summers. The species that were adapted to the soil and site characteristics—tall fescue, reed canarygrass, and alsike clover—were not acceptable to the owner. A large investment was made in land and facilities that could never produce the desired outcome.

The foremost consideration before purchasing the land should have been the soil and site characteristics and suitability of the land for the grass and legume species the owner desired for pasture. That thought never crossed the mind of the owner until the land had been cleared and construction of buildings, fences, and facilities was completed—a costly mistake. Unfortunately, this example is not an unusual occurrence, particularly with small and part-time farmers.

This is one of the reasons so much emphasis is placed on conducting a thorough and complete inventory of all available resources that will be used in the pasture and grazing program. Among the questions to be addressed in the process of selecting adapted grass and legume species are:

- What are the soil limitations of each field in the grazing system?
- Is drainage a limiting factor in any of the grazing fields? Poorly drained soils stress plant root systems. Species differ in their ability to persist on poorly drained soils.
- Are fertility and pH limiting factors? It is important to know not only what the fertility and pH limitations are but also to know where they are (which fields). Old, permanent pastures typically have low pH and fertility, severe limitations especially for legume production. Soil pH and fertility are correctable limitations, but keep in mind that it may take 2–3 years or more for surface applications of lime and fertilizer to effectively change levels in the root zone.
- Are rooting depth and topography limiting in any of the grazing fields? Shallow soils are droughty, which will stress plants during hot, dry weather. Steep slopes limit access and operation of equipment for liming, fertilizing, clipping, etc.

MATCHING PLANTS TO THE GRAZING SYSTEM GOALS AND OBJECTIVES

Once the list of species adapted to the soil conditions within the grazing system has been determined, further decisions on forage selections must be made with the "end user"—the grazing animal—in mind. One consideration should be the nutritional needs of the species and classes of livestock that are to be grazed on each pasture field. Will the pasture support the type of animals that will be grazing? Will it be used during early lactation and breeding (a time of maximum need for both quality and quantity)? Is it free of antiquality constituents such as the toxic type of endophyte that can occur in tall fescue? Will it be for growth of replacement animals or for backgrounding feeder calves? Growing animals and lactating animals require high-quality forage (protein and energy) to meet production requirements.

What role will each pasture field play? Will it be part of the "backbone" of the grazing system or will it be a supplemental forage to fill a gap or low point in forage production? What will be the primary season of use? For example, tall fescue selected for fall or winter grazing will not provide grazing in August, September, and October. Likewise, a productive summer grass such as switchgrass will have a relatively short (but productive) growing season compared to tall fescue or orchardgrass.

What will be the frequency of grazing and the length of the rest periods? Kentucky bluegrass and white clover are more likely to persist under close grazing on a somewhat poorly drained soil with high fertility than are orchardgrass and red clover. What is the extent of traffic? Is the primary intended purpose of the pasture to be an exercise lot for horses or a loafing lot for dairy cows? In these cases an endophyte-infected tall fescue may be the species of choice.

It is helpful to have some understanding of the forage quality of various species of forage grasses and legumes and to choose those that will best meet the nutritional requirements of the species and classes of livestock being grazed. Consider having fields of several different grass and legume mixtures to provide forage in as many months of the year as possible, thereby reducing stored feed costs. Yields of mixtures are usually higher than when either is grown alone. Legumes increase the protein content and often the palatability of the forage. With 25–35% or more legume plants in a mixture, N fertilizer is not necessary. Grasses are less susceptible than legumes to winter injury and frost heaving and offer protection to legumes in the stand. Legume-grass stands are generally productive longer than pure legume stands, give better erosion control, and have less chance of weed infestation. Field curing of hay is aided by having grass in the mixture. Mixtures also provide greater flexibility of use.

Mixtures do have some disadvantages. A higher level of management is required to maintain a proper balance of legume and grass. Herbicide choices for weed control are limited because few herbicides are selective for both legumes and grasses. The growth patterns of the legume and grass may not be compatible, so they may reach recommended harvest stages at different times.

Prepackaged "shotgun" mixtures of numerous grasses and legumes usually have no advantage over simpler mixtures of one or two grasses and one or two legumes carefully designed to match specific species to the soil and site characteristics and grazing system goals. Species in a mixture should have similar maturity dates, compatible growth characteristics, and be adapted to the intended use. Species that germinate quickly and have vigorous seedling development ensure early production, but too much of that species (e.g., more than 5 pounds/acre of annual or perennial ryegrass) can crowd out others. Species in a mixture should also be similar in palatability; otherwise, unpalatable species will soon dominate from lack of grazing.

High-quality seed is essential for good, persistent forage stands; thus, certified seed of known varieties should be used. Certified seed carries a label certifying that it is the seed of the particular variety listed on the label, that it meets minimum standards of quality in purity and germination, and that it has low weed seed content (usually less than 0.25%). There are no standards for noncertified seed other than state limits on weed seed content, which can be as high as 2%. Newer varieties provide superior agronomic characteristics that economically justify their selection over older, lower cost varieties that have less disease and insect resistance and lower yield capabilities and often produce big disappointments.

ESTABLISHING NEW STANDS

Common terminology used to describe forage seeding methods includes conventional, no-till, broadcast, drilled, cultipacker, frost, and walk-in or tread-in. Two or more of these terms might describe a seeding. For example, conventional and no-till seedings might be broadcast-seeded or drill-seeded. Seeding methods will be categorized here by the type of seedbed—tilled versus no-till. The erosion potential of a field must be considered before choosing a tillage method. Primary tillage implements such as the moldboard plow, chisel plow, and heavy disks bury much or all of the surface residue, leaving bare soil subject to runoff and erosion, especially on sloping fields.

Tilled Seedbed Seedings

Tilled seedbed seedings are sometimes referred to as conventional seedings because conventional tillage practices (plowing, disking, harrowing, etc.) are used to prepare the seedbed. The purposes of tillage are to loosen the soil, eliminate existing vegetation, turn under surface weed seeds, incorporate lime and fertilizer into the soil, and provide a smooth surface for harvesting operations. A satisfactory tillage sequence controls weeds and provides a firm seedbed with just enough loose surface soil for shallow seed placement with good seed-soil contact.

Tillage that leaves some residue on the surface will provide better conditions for developing seedlings than will an overworked seedbed with no residue or mulch. However, too much surface residue or trash can result in seed placement that is too shallow because seeding units ride on top of the residue. Cloddy or trashy seedbeds are usually too rough or uneven for uniform depth control and seed placement and are too coarse for good seed-soil contact. Overworking the soil (too much tillage) results in fluffy, powdery seedbeds that dry out quickly and may be too fine, increasing the potential for surface crusting following rainfall and for poor seedling emergence. Crusting is particularly a problem with small-seeded legumes. Small clods or soil aggregates can be beneficial to prevent soil crusting. The primary problems with conventionally tilled seedbed seedings are soil moisture loss during tillage and the soil erosion potential until the seeding becomes established.

The three most common methods of seeding on tilled seedbeds are cultipacker seeding, drill seeding, and broadcast frost-seeding. Cultipacker seeders consist of two sets of corrugated rollers with seed-metering boxes or hoppers that drop the seed between the two sets of rollers. The first set of rollers firms the soil into shallow corrugations, then drops seed. The second set of rollers splits the ridges of the corrugations, covering the seed and firming the soil over and around it. Cultipacker seeders provide optimum seed placement and good seed-soil contact if seedbeds are properly prepared.

On medium- and heavy-textured soils, some of the seed remains on the top and sides of the ridges as well as at the bottom of the corrugations. Since these corrugations are split by the second set of rollers, the seed is distributed across a range of depths from ¼ to 1 inch. On sandy soils, most of the seed falls to the bottom of the corrugations and deeper coverage is obtained as is desired. Cultipacker seeders should not be used on heavy soils having a moist surface because crusting is likely. Adequate soil coverage may not be obtained with a cultipacker seeder on seedbeds having heavy crop residues, clods, or stones or with light, fluffy grass seeds such as smooth bromegrass.

Variations of cultipacker seeding include aerial, fluid (or suspension), and broadcast seeding. The ground is cultipacked, seed is distributed, and cultipacking is repeated. Distributing seed through sprayers, which is referred to as hydro-, fluid, or suspension seeding, is a very effective way of broadcasting seed uniformly over large areas in a short time, usually by custom applicators.

Grain drills with grass and legume seed attachments and seed tubes extending to the ground can accurately meter the seed, but controlling the depth of seeding can be difficult. Seed may be covered too deeply, especially if the disks (shovels or hoes on older drills) are set too deep and the seed is dropped near the opener. Seed that falls beneath soil thrown up by the openers frequently is covered with too much soil for the seedlings to emerge. Drills with press wheels generally provide excellent results if a uniform shallow depth can be maintained. They also work better than cultipacker seeders on fields with crop residue.

Some producers still broadcast seed in late winter on the soil surface over fall-sown small grains. Before the advent of no-till drills this was the common method for seeding alfalfa and other hay crops. Freezing and thawing action (honeycombing of the soil surface with ice crystals) plus rain will cover seed with soil to about the right depth. Frost-seeding is successful only during relatively short periods when soil and climate conditions are right.

No-Till Seedings

No-till seeding reduces soil erosion and conserves soil moisture for germination and seedling growth. Additional benefits are reduced fuel, labor, and time requirements.

Seedings can be made into a variety of ground covers. In areas where winter cereals are grown, a late summer seeding into winter barley or wheat stubble is one of the most reliable methods and one that has great flexibility in terms of matching seeding dates, soils, and weather conditions. Following harvest of the grain crop, allow time for as many weed seeds to germinate as practical so that maximum weed kill will be accomplished with application of a nonselective herbicide such as paraquat or glyphosate. This will generally be 2–3 weeks following grain harvest. A second herbicide application should be made at seeding after a second flush of weeds has germinated. Because soil tillage is not required, seedings can be made much sooner following rain than with conventional tillage and more soil moisture will remain for the new seeding than with a conventionally tilled seedbed. The stubble reduces soil erosion and provides protection for the seedlings from intense sunlight and damage from blowing soil particles.

When perennial broadleaf weeds are present, selective translocated herbicides such as 2,4-D and dicamba will provide control, but treatment must be at least 30 days prior to seeding, and 6 months is better (4). Growing row crops such as corn or small grains on weedy pastures or hay fields for a year or two can be beneficial for controlling problem weeds prior to seeding new pasture or hay stands. More effective weed control is possible with row crops, because alternative herbicides can be used that can not be used on hay or pasture. Growing row crops also helps to break down the sod and reduce possible insect, slug, snail, and disease problems. Crops used in this way are often referred to as "break crops," because they provide a break in an otherwise continuous crop.

Summer annual grasses such as sudangrass, sorghum-sudangrass hybrids, and millets can also be used as break crops. These summer annual grasses grow vigorously and weeds are generally not able to compete. Allelopathic toxins from these grasses may also inhibit weed germination and development. Thus, they are good smother crops in preparation for forage seedings. As with winter wheat or barley, no-till seedings can be made into the killed stubble. However, since the summer annual grasses do grow vigorously during the summer, they can deplete soil moisture in dry years if not killed 3–6 weeks prior to the forage seeding.

Brassicas such as kale, rape, swedes, turnips, and tyfon can also be used as break crops between old and new pasture stands. However, these crops do not grow as vigorously as the summer annual grasses and are usually not as effective in smothering weeds.

Where both broadleaf and grass perennials are present, a nonselective translocated herbicide such as glyphosate should be used to kill all of the vegetation present. In this case, delaying seeding for 1–3 weeks after application will give better weed control (12, 22).

NO-TILL PASTURE RENOVATION

Much of the permanent pasture in the humid United States is not as productive as it could be. This is due mainly to the lack of adequate fertilization, poor grazing management, and poor yield potential and seasonal distribution of growth of the forage species present. Renovation of these pastures could improve production more than twofold, depending on soil characteristics and the condition of existing sod (5–7, 15, 18).

"Renovation," as used here, is the improvement of a pasture by the partial or complete destruction of the sod, plus liming, fertilizing, weed control, and seeding as may be required to establish desired forage plants without an intervening crop. No-till technology enables farmers to renovate old pastures without plowing or disking. Herbicides such as paraquat and glyphosate enable suppression of existing vegetation without tillage. No-till drills can place seeds at the proper depth and ensure good seed-soil contact.

Suppressing Competition from Existing Vegetation

Pastures to be sod-seeded should be very closely grazed or clipped 4–6 weeks prior to a late summer seeding or during the fall/early winter prior to a late winter seeding. This will reduce top growth and surface residue, assisting in suppressing competition from existing vegetation. This is the one time in pasture management when overgrazing is beneficial. Grazing is preferred to clipping because animals can graze closer to the ground than machines can clip and more residue is removed. Following grazing or clipping, plants should be allowed to green up for the herbicides to be effective. There are two methods for applying the herbicide—band and broadcast. The method to be used depends on the composition and condition of the existing sod, whether the renovation is a completely new seeding or simply the introduction of legume(s), and the time of year of the seeding.

Band Application of Contact Herbicide for Late Winter Seeding

If an adequate stand of productive grasses such as orchardgrass, tall fescue, or smooth bromegrass is present and the renovation is being made in late winter/early spring primarily to introduce legumes, then a band application of contact herbicide such as paraquat at seeding is recommended. Band application of herbicide entails applying a band of spray over the seed rows so that about half of the surface area is sprayed, for example, 5-inch bands over 10-inch row spacing. Unsprayed bands are left between the rows. This works best with 8–10-inch row spacing drills. The spray boom is mounted on the drill, generally ahead of the seeding units. The contact herbicide will suppress the grass within the band for 4–6 weeks, reducing the amount of competition for the legume seedlings. The grass will then recover. If productive species are not present in the existing sod and both new grass and legume species are being seeded, then broadcast application of a nonselective systemic herbicide such as glyphosate is recommended to kill all vegetation.

Band application of the herbicide has several advantages. First, band application generally results in fewer weeds in the new seeding. Whenever plant growth is suppressed or destroyed, weed seeds in the soil germinate. This is especially true for broadleaf weeds and summer annual grasses that are common problems in spring seedings. Band spraying allows about half of the sod to remain actively growing, thus helping to retard weed growth. Second, the remaining sod provides pasture for grazing much earlier than when all of the sod is suppressed or killed with a broadcast application of herbicide. Since half of the original sward remains intact, pastures can produce significant grazing material 30–60 days after seeding with band application of herbicide (if label restrictions allow). Wait at least 30 days between contact herbicide application and grazing. Even though the spring flush of growth is reduced by about half with the banded herbicide, total production for the seeding year is not reduced because the seeded legumes provide increased yields of higher quality forage later in the season compared to the untreated sward (1, 17). Animal gains in the seeding year are almost double those of unimproved swards (5). Third, if productive desirable species are present, there is no need to kill them and reseed the same species again. Fourth, there may be less insect feeding on new seedlings since there is other vegetation for them to feed on.

Banded contact herbicide should be applied within recommended rates to that half of the area actually sprayed. This means that for each acre of pasture being renovated, only half of the normal volume of material will be used.

Broadcast Application of Herbicide for Late Summer Seeding

For late summer seedings, it is important that all existing vegetation be eliminated 4–6 weeks prior to seeding so that it will not deplete soil moisture needed by the new seeding or compete with the new seedlings during establishment. The decision of which herbicide to use should be based on the type of vegetation present. If the pasture has been closely grazed or mowed, allow adequate regrowth before spraying to obtain effective control. The initial application can be followed 4–6 weeks later by an application of a contact herbicide at the time of seeding for control of weeds that germinated after the first application.

Late Winter Frost-Seedings

An alternative method of no-till seeding is frost-seeding in late winter before grass breaks dormancy and initiates spring growth. As with tilled seedbed frost seedings, the seed is broadcast when alternate freezing and thawing of the soil surface is occurring. This works best where grass stands are thin and where the grass sward was heavily grazed the previous fall and early winter. On well-drained soils, where puddling is not a problem, the trampling effect of animals on the pasture can improve seed-soil contact, but animals should be removed when seed germination begins. Frost-seeding generally works quite well with red and white clovers and often with grasses such as annual and perennial ryegrass that develop vigorously following germination, but is usually much less successful with alfalfa, birdsfoot trefoil and grasses such as orchardgrass, tall fescue, timothy, etc.

NO-TILL SEEDING EQUIPMENT

No-till drills should be heavy enough to ensure proper thatch and soil penetration or be designed for the addition of supplemental weight. They should also have sufficient durability for no-till operating conditions. Desirable drill features that make seedings more reliable include:

- rolling coulters to cut through mulch covers (plain, notched, or ripple coulters will cut and penetrate better than wide 1- or 2-inch fluted coulters),

- double-disk seed furrow openers that line up precisely with the cutting coulters for proper seed placement in the soil,
- depth bands, wheels, or other methods of controlling the depth of each seeding unit,
- independently operating seed placement units to follow the soil terrain, and
- a press wheel behind each seeding unit to firm the soil around the seed.

Narrow press wheels, no wider than 1 inch, are preferred unless the press wheels have a narrow center ridge running in the slit left by the double-disk openers. The press wheel increases in importance as soil texture increases and soil moisture decreases. However, with heavy, wet soil conditions, press wheels can be detrimental. In this case it may be desirable to remove the press wheels and leave the seed furrow open for the seed to germinate and develop at the bottom of the open slit. If the slit is closed with a press wheel, the seed may be covered too deeply or the soil may be compacted and the seedling may be unable to emerge.

WHEN TO SEED

The primary seeding periods for cool-season species are late winter through early spring (late February to mid-May, depending upon location) and late summer (August to mid-September). Late winter through early spring seedings are most common. Soil moisture and rainfall are generally good, evaporation is less, and soil moisture is retained longer during the establishment period than with late summer seedings. However, seeding too early in cold, wet soils can result in poor germination, seedling loss due to fungal diseases, and weak stands. Late spring seedings often fail, however, due to stress from high temperature and lack of moisture. The surface 1–2 inches of soil can dry quickly in the summer and young seedlings desiccate. Also, annual weeds are more of a problem with late spring seedings.

Late winter through early spring is the preferred time for seeding legumes into an existing productive stand of grass. These seedings should be made as early as possible to minimize grass and weed competition. Early seedings tend to be more competitive, less weedy, have reduced effects of slugs (3; also see chapter 6), and result in better stands than seedings made late in the spring. The "best" time to seed a new seeding largely depends on the local climate, which varies with altitude and latitude across the region.

Late summer is the preferred time for seeding if the existing species are to be eliminated and both the grass and legume must be seeded. Soil moisture can frequently be a limiting factor with late summer seedings. This is especially true for band applications of herbicide, and therefore it is not recommended for late summer seedings. The existing sod may deplete soil moisture for germination and seedling development. Seeding should be timed with rainfall so that adequate soil moisture is available at the time of seeding.

Advantages of Late Summer Seedings

- There is less competition from weeds—herbicides may not be needed.
- Seedings can be made after early harvested crops such as winter wheat and barley and will be fully productive the following year.
- The spring work load is lessened.
- Liming, fertilization, and tillage are done during drier weather, thus reducing the risk of soil compaction.
- Damping-off diseases (*Pythium* spp. and *Rhizoctonia* spp.) are not usually a problem.

Late summer seedings need sufficient time, moisture, and sunlight for adequate growth before a killing frost. These seedings should be made early enough to allow at least 6 weeks for growth after germination and emergence. Seedlings should be at least 3–4 inches tall before a killing frost. Seedings made after seasonal cutoff dates are more subject to winter injury and possible winter killing because the plants do not have as much time to develop and become established. Although damping-off diseases are not usually a problem, late summer legume seedings are more susceptible to Sclerotinia crown and stem rot than are late winter to early spring seedings. New seedings can be completely destroyed when conditions are favorable for this disease development. Seedings should be made at the earliest possible date so that seedlings are well established by the time infection occurs.

SEEDING DEPTH

Seeding depth varies with soil type (sand, clay, or loam), soil moisture availability, time of seeding, and firmness of the seedbed. Since small-seeded forage grasses and legumes have a very small supply of stored energy to support the developing seedling, proper seeding depth is important in getting good stands. Seeds placed too deep are not likely to emerge. Seeds placed on the surface or at a very shallow depth or in a loose or cloddy seedbed often do not have adequate seed-soil contact. In these cases, dry soil conditions following seeding usually result in desiccation and death of the seedlings. Thus a firm seedbed is essential for proper seed placement, good seed-soil contact, and successful establishment.

The number of times a tilled seedbed needs to be cultipacked depends on the type of seeder being used, the soil type, and the amount of soil moisture. For example, under dry soil conditions with medium- and heavy-textured soils, cultipacking once prior to seeding with a cultipacker-type seeder may be sufficient. However, if seeding with a drill, cultipacking twice prior to seeding may be necessary. A fluid or suspension seeding may require cultipacking once before and twice after seeding. Lighter, sandy-textured soils usually require fewer trips across the field. One measure that a seedbed is properly firm is that your footprint should sink no deeper than ⅜–½ inch into the soil.

Seed should be covered with enough soil to provide moist conditions for germination. Under humid conditions, best results are obtained when seed placement is between ¼ and ½ inch deep. Under more arid conditions, such as in sandy soils, or to reach moist soil, the seed must be sown deeper. However, seed placed deeper than 1 inch may not emerge or be so weakened that survival is reduced. Generally, the optimum seeding depths are ¼–½ inch on clay and loam soils and ½–1 inch on sandy soils. Shallower depths within these ranges are better for species with smaller seed sizes and for early spring seedings—moisture is usually abundant and soil temperature is warmer near the surface. Deeper depths are recommended for species with larger seed sizes and for late spring and summer seedings when moisture conditions are less favorable. It is not unusual to see a few seeds on the soil surface, especially when seeding at the shallower depths.

SEEDING RATES

Recommended seeding rates vary from state to state, and in some cases within states, due to differences in soils, climate, and establishment methods. Pasture seeding rates are higher than hay seeding rates to provide a denser sod for grazing. This is especially true for horse pastures. Selecting specific seeding rates within the recommended ranges depends on factors such as:

- Soil types and fertility. Lower rates can be used when seeding on light, sandy soils because seedling emergence is easier than in heavy soils. Higher rates should be used on lower fertility soils because more seedling losses are likely to occur and plants will be smaller and less vigorous, with less spread of the crowns.

- Amount and distribution of rainfall. In areas of limited rainfall and on light, sandy soils, lower seeding rates are used because soil moisture may not be sufficient to support dense plant populations due to competition for moisture. Also, seedling losses due to damping-off diseases are generally less under these conditions.
- Condition of the seedbed and seeding method. Lower rates can be used when seeding into well-prepared seedbeds where uniform seed coverage and good seed-soil contact can be achieved. Higher rates should be used when seedbed preparation and seeding techniques are less than optimum.
- Seed quality. Higher rates are necessary if the seed lot has low germinability, a high percentage of hard seed, or a low percentage of pure seed. Use of certified seed helps to ensure variety purity and high seed quality. Pure live seed is a term used to express the quality of seeds. It is expressed as a percentage of a seed lot that is pure seed that will germinate. The percentage is determined by multiplying the percentage of pure seed by the percentage of germination and dividing by 100.
- The desired composition of the stand if seeding a legume-grass mixture.
- Whether seeding with or without a companion crop. Higher rates are recommended when seeding without a companion crop because more plants are needed to obtain high yields in the seeding year and to reduce weed invasion. It is common for no more than a third of the sown seed to produce seedlings and only half of those to survive the first year (16, 20). Seedling survival frequently is as low as 20% of all seedlings that germinate (2).

Table 7-1 (pp. 112–113) lists seeding rate recommendations for various choices of species and mixtures based on soil type and primary use. Seeding rates do not need to be increased for no-till seedings.

To achieve accurate seeding rates, seeders must be properly calibrated. Seeders should be calibrated for each different lot of seed and each year, even if the same variety is being planted. Different lots of seed within the same variety can have different seed size and test weight. For example, alfalfa seeding rates can vary by 5–6 pounds/acre with the same seeder setting (19). Even though seed size is larger and will weigh almost one-third more than uncoated seed, lime-coated seed often flows though a seeder 20–30% faster than uncoated seed. This is opposite of what you might expect. If you don't want to take the time and effort to follow the calibration procedure for the seeder, at least determine the acreage seeded with the first half bag of seed, estimate the seeding rate, and adjust the seeder setting accordingly.

INOCULATION OF LEGUME SEEDS

All legume seed should be inoculated with the proper strains of N-fixing bacteria *(Rhizobium* spp.) before seeding. If the seed is not preinoculated or the seeding date is beyond the expiration date for the inoculant on preinoculated seed, the seed should be inoculated with a fresh culture prior to seeding. A specific relationship exists between bacteria and the host plant, so be sure that the inoculant is specific for the legume species being seeded.

Common on-farm inoculation techniques such as the use of water, milk, and colas for making the surface of the seed sticky for adherence of inoculant to the seed are often ineffective. Much of the inoculant initially adhering to the seed sloughs off when the seed dries. The use of colas

Table 7-1. Recommended hay and pasture seeding rates for new stands.

Species and mixture recommendations according to soil type and primary use	Seeding rate (lb/ac)
I. Deep, well-drained fertile soils	
A. Primarily for hay, silage, or greenchop	
Alfalfa alone	**15–20**
Alfalfa and	**8–12**
Late-maturing orchardgrass ***or***	2–6
Timothy ***or***	4–6
Smooth bromegrass	6–8
Alfalfa and	**8–10**
Reed canarygrass	8–10
Red clover alone	**10–12**
Red clover and	**6–8**
Timothy ***or***	4–6
Late-maturing orchardgrass	4–6
Birdsfoot trefoil alone	**7–10**
(Upright or European-type varieties–Cascade, Viking, Leo, etc.)	
Birdsfoot trefoil and	**5–8**
Timothy ***or***	4–6
Late-maturing orchardgrass ***or***	2–4
Reed canarygrass	8–10
Orchardgrass alone	**10**
Reed canarygrass alone	**12–14**
Smooth bromegrass alone	**12–15**
Timothy alone	**8–10**
B. Primarily for pasture	
Orchardgrass and	**8–10**
Alfalfa ***or***	8–10
Red clover + white (ladino-type) clover ***or***	6–8 + 1–2
Birdsfoot trefoil (prostrate and semierect Empire-type varieties– AU-Dewey, Carroll, Dawn, Fergus, Norcen, etc.)	5–8
Orchardgrass + perennial ryegrass and	**8–10 + 3–5**
Red clover + white (ladino-type) clover	6–8 +1–2
Orchardgrass + Kentucky bluegrass and	**8–12 + 2–4**
White (ladino-type) clover (especially suited for horse pasture)	1
Tall fescue and	**10–15**
Alfalfa ***or***	8–10
Red clover + white (ladino-type) clover ***or***	6–8 + 1–2
Birdsfoot trefoil	5–8
Smooth bromegrass and	**8–10**
Alfalfa ***or***	8–10
Red clover + white (ladino-type) clover ***or***	6–8 + 1–2
Birdsfoot trefoil	5–8
II. Wet or poorly drained soils	
A. Primarily for hay, silage, or greenchop	
Red clover + alsike clover and	**6–8 + 2–4**
Timothy ***or***	4–6
Reed canarygrass	8–10
Birdsfoot trefoil and	**5–8**
Timothy or	4–6
Reed canarygrass or	8–10
Late-maturing orchardgrass	2–4
Birdsfoot trefoil alone	**7–10**
Reed canarygrass alone	**12–14**

Table 7-1 (cont.). Recommended hay and pasture seeding rates for new stands.

Species and mixture recommendations according to soil type and primary use	Seeding rate (lb/ac)
B. Primarily for pasture	
Tall fescue and	**10–15**
Birdsfoot trefoil ***or***	5–8
Alsike + white (ladino-type) clover	2–4 + 1–2
Reed canarygrass and	**8–10**
Birdsfoot trefoil ***or***	5–8
Alsike + white (ladino-type) clover	2–4 + 1–2
Tall fescue alone	**15–20**
Reed canarygrass alone	**12–14**
III. Droughty soils: Sandy, shallow, steep, shaly, or gravely	
A. Primarily for hay, silage, or greenchop	
Alfalfa alone	**15–20**
Alfalfa and	**8–12**
Late-maturing orchardgrass ***or***	2–6
Smooth bromegrass ***or***	6–8
Alfalfa and	**8–10**
Reed canarygrass	8–10
Red clover alone	**10–12**
Red clover and	**6–8**
Late-maturing orchardgrass ***or***	4–6
Reed canarygrass	8–10
Birdsfoot trefoil alone	**7–10**
Birdsfoot trefoil and	**5–8**
Late-maturing orchardgrass ***or***	2–4
Reed canarygrass	8–10
B. Primarily for pasture	
Tall fescue and	**10–15**
Alfalfa ***or***	8–10
Red clover ***or***	6–8
Birdsfoot trefoil	5–8
Orchardgrass and	**8–10**
Alfalfa ***or***	8–10
Red clover ***or***	6–8
Birdsfoot trefoil	5–8
Perennial ryegrass and	**12–15**
Alfalfa ***or***	8–10
Red clover ***or***	6–8
Birdsfoot trefoil	5–8
Switchgrass	**8–10 (PLS)[a]**
Big bluestem	**8–10 (PLS)**
Caucasian bluestem	**3–4 (PLS)**
Eastern gamagrass	**8 (PLS)**
Indiangrass	**6–8 (PLS) drilled** **12–15 (PLS) broadcast**

[a] *PLS = pure live seed = (% germination x % pure seed) ÷ 100.*

can be detrimental to the rhizobia because these bacteria are sensitive to acidic conditions, and colas are acidic. Dry or drill box treatments are also usually ineffective and result in poor inoculum adhesion. Successful legume inoculation under adverse conditions generally depends on coating large amounts of inoculant on the seed, keeping it there until the seed is in the ground, and ensuring survival of the rhizobia in the soil until the young seedlings are infected. More rhizobia in the area of the developing root means more nodules and thus more N fixed.

A sugar-water solution can be used as an effective sticking agent at an economical cost. Use at least 2 cups of sugar per quart of water, thoroughly moisten the seed, and then add 16 ounces of inoculum per 20 pounds of seed. This is much more inoculum than most growers are accustomed to applying, but the purpose is to increase the number of rhizobia per seed so that enough will survive for effective nodulation even under adverse conditions. Cornstarch can be added after thoroughly mixing the inoculant with the seed. Add cornstarch until the seed is dry and will flow well through the seeder so seeding can begin immediately. It also provides an energy source for the rhizobia.

Heat, direct sunlight, and drying are detrimental to survival of the rhizobial inoculant. Seed dealers should store inoculant in a refrigerator, and it should be kept in a refrigerator from the time of purchase to the time of use. Always check the expiration date before purchasing any inoculant materials or any preinoculated seed. Do not purchase out-of-date inoculant or seed.

MANAGEMENT DURING ESTABLISHMENT

Weeds often invade new seedings, and the stand may be reduced if they are not controlled. Clipping may be necessary, but it should not be done too early. If clipped too early, only the tops of the weeds will be removed, leaving active buds on the stubble to produce new branches and even more competition. Sufficient weed growth should be allowed so that most active buds are removed when the new seeding is clipped. Most forage grasses and legumes regrow from crown buds and are usually not seriously damaged by cutting. However, clipping too frequently can reduce seedling development as well as forage yields the following year (16).

If a band application of herbicide was used in pasture renovation, the untreated bands may compete with the young seedlings if not grazed or mowed. One of the advantages of band application previously mentioned was to allow earlier grazing on the seeded area. Be sure to follow herbicide label restrictions before grazing. These pastures can be grazed down until the animals start eating the leaves and shoots of the young seedlings. At that point the animals need to be removed to allow continued growth of the seedlings. Do not regraze until seedlings are at least 3 inches tall.

Even though late summer seedings may make considerable growth during the late summer and fall period, they should not be cut or grazed. Clipping or grazing seedling stands weakens the plants and results in greater susceptibility to winter killing. Annual and perennial ryegrass, and possibly festulolium, are exceptions to this rule because they germinate quickly and grow rapidly. Excess growth of these species may contribute to winter killing. Grazing or clipping of new spring seedings should end 4–6 weeks prior to the average killing frost date to allow a buildup of reserves for winter. Avoid grazing during wet periods, especially on tilled seedbeds.

The new seedings should be monitored for slugs, insects, and diseases at least weekly for the first 6–8 weeks. Producers often fail to monitor new seedings, and it is difficult, if not impossible, to determine the cause of seeding failures when you don't know what happened to the seedlings during that 6–8-week period following seeding.

SUMMARY–MANAGE, MANAGE, MANAGE

There are no shortcuts or substitutes for good management practices and procedures for establishing forages. Poor management before and after seeding can result in wasted effort. It is often surprising just how productive some old, worn-out pastures and hay fields can become. But once this initial improvement has been made, there is too often a tendency for producers to slip back into old practices of overgrazing and inadequate fertilization, and soon the new seedings are back to weeds and low productivity. After renovation, it is necessary to follow up with good forage management to realize continuing benefits from a new seeding.

Appendices

APPENDIX A. IDENTIFICATION OF MAJOR FORAGE SPECIES

Common name/ scientific name [a]	Life cycle	Season of growth	ID key	Growth habit/ root structure	Ease of establishment (1–5) 1=easy	Harvest [b] Height (stubble) in.	Harvest [b] Frequency (weeks)	Life expectancy (years) [c]	Regrowth potential (fair, good, excellent)
GRASSES									
Kentucky bluegrass *(Poa pratensis)*	Perennial	Cool-season	Leaf folded in the sheath; leaf blade with boat-shaped tip; auricle well developed; ligule small-toothed, truncated	Rhizomes	3	2–3	2–4	10+	Fair
Orchardgrass *(Dactylis glomerata)*	Perennial	Cool-season	Leaf folded in the sheath; leaf blade strongly folded; no auricle; ligule pointed, jagged	Bunch	1	2–4	2–4	10+	Excellent
Reed canarygrass *(Phalaris arundinacea)*	Perennial	Cool-season	Leaf rolled in the sheath; leaf blade flat with rough margin; no auricle; ligule large, acute	Rhizomes	3–4	2–4	2–4	10+	Excellent
Rescuegrass *(Bromus cartharticus)*	Perennial acts as annual	Cool-season	Leaf rolled in the bud-shoot; leaf appears flat or folded; sheath densely covered with fine hairs; no auricle; long ligule, membranous, and fringed	Bunch	2	3–4	4–6	2–4	Good
Ryegrass *(Lolium perenne)*	Perennial	Cool-season	Leaf folded in the sheath; leaf blade flat; auricle small, clawlike; ligule small, blunt to pointed	Bunch	1	2–4	2–4	2–3	Excellent
Smooth bromegrass *(Bromus inermis)*	Perennial	Cool-season	Leaf rolled in the sheath; leaf blade veins distorted to form "M" toward tip end; no auricle; ligule membranous	Rhizomes	3	3–4	4–5	10+	Good
Tall fescue *(Festuca arundinacea)*	Perennial	Cool-season	Leaf rolled in the sheath; leaf blade with obvious parallel veins; rough along the edges; no auricle; ligule blunt	Bunch, short rhizomes	2	2–4	2–4	4–20+	Excellent
Timothy *(Phleum pratense)*	Perennial	Cool-season	Leaf rolled in the sheath; leaf blade wide, twisted, ribbed with rough margin; no auricle; ligule acute, toothed	Bunch	3	2–4	2–4	10+	Fair
Barley *(Hordeum vulgare)*	Annual	Cool-season	Leaf rolled in the bud; long, clasping auricles, clawlike, glabrous; auricles barely larger than those of wheat or rye; short or truncated ligule	Bunch	1	2–4	2–5	1	Excellent

(APPENDIX A. IDENTIFICATION OF MAJOR FORAGE SPECIES, cont.)

Common name/ scientific name [a]	Life cycle	Season of growth	ID key	Growth habit/ root structure	Ease of establishment (1-5) 1=easy	Harvest [b] Height (stubble) in.	Frequency (weeks)	Life expectancy (years) [c]	Regrowth potential (fair, good, excellent)
Cheat/chess *(Bromus secalinus)*	Annual	Cool-season	Leaf rolled in the sheath; leaf blade broad, smooth; auricle absent; ligule membranous, toothed	Bunch	–	–	–	1	Fair
Downy chess/ downy bromegrass/ cheatgrass *(Bromus tectorum)*	Annual	Cool-season	Leaf rolled in the sheath; leaf blade flat, covered with short, stiff hairs, white midrib; auricle absent; ligule membranous, jagged; dense fringe of white hairs	Bunch	–	–	–	1	Fair
Oats *(Avena sativa)*	Annual	Cool-season	Leaf rolled in the sheath; leaf blade flat, wide; auricle absent; ligule membranous	Bunch	1	2–4	2–4	1	Fair
Rye *(Secale cereale)*	Annual	Cool-season	Leaf rolled in the bud-shoot; small or no auricle; ligule membranous, rounded, jagged, may be ciliate	Bunch	1	2–4	2–4	1	Excellent
Ryegrass (Italian) *(Lolium multiflorum)*	Annual	Cool-season	Leaf rolled in the sheath; leaf blade flat; auricle small, clawlike; ligule small, blunt to pointed	Bunch	1	2–4	2–4	1	Excellent
Wheat *(Triticum aestivum)*	Annual	Cool-season	Leaf rolled in the sheath; leaf blade flat; auricle clawlike; ligule membranous, blunt	Bunch	1	2–2	2–4	1	Excellent
Bermudagrass *(Cynodon dactylon)*	Perennial	Warm-season	Leaf folded in the sheath; leaf blade flat or V-shaped, lower blade covered with silky hairs; auricle absent; ligule membranous	Rhizomes and stolons	1	1–3	2–4	20+	Excellent
Big bluestem *(Andropogon gerardi)*	Perennial	Warm-season	Leaf rolled in the bud-shoot; young shoots, somewhat flattened at the base; base of lower sheath covered with silky hair; auricle absent; ligule membranous, sometimes fringed	Bunch with or without rhizomes	3	6–8	4–6	20+	Good
Caucasian bluestem *(Andropogon caucasius)*	Perennial	Warm-season	Leaf rolled in the sheath; leaf blade narrow, hairy; auricle absent; ligule hairy	Bunch	4	3–4	4–6	10+	Excellent
Dallisgrass *(Paspalum dilatatum)*	Perennial	Warm-season	Leaf rolled in the bud; leaf blades flat; leaf margins with fine hair; collar broad with a few long silky hairs; auricle absent; ligule tall and membranous	Short rhizomes	5	2–4	2–4	10+	Good

(APPENDIX A. IDENTIFICATION OF MAJOR FORAGE SPECIES, cont.)

Common name/ scientific name [a]	Life cycle	Season of growth	ID key	Growth habit/ root structure	Ease of establishment (1-5) 1=easy	Harvest [b] Height (stubble) in.	Frequency (weeks)	Life expectancy (years) [c]	Regrowth potential (fair, good, excellent)
Eastern gamagrass *(Tripsacum dactyloides)*	Perennial	Warm-season	Leaf rolled in the sheath; leaf blade wide with rough and sharp margins	Scaly rhizomes	5	6–8	2–5	10+	Excellent
Indiangrass *(Sorghastrum nutans)*	Perennial	Warm-season	Leaf rolled in the sheath; leaf blade thick, flat, light-colored midrib, smooth; auricle small to absent; ligule long, hairy fringe	Rhizomes	3	6–8	2–5	10+	Excellent
Johnsongrass *(Sorghum halepense)*	Perennial	Warm-season	Leaf rolled in the sheath; leaf blade smooth; auricle absent; ligule long, prominent membrane with fringe of hairs	Thick rhizomes	1	6–8	2–5	20+	Excellent
Switchgrass *(Panicum virgatum)*	Perennial	Warm-season	Leaf rolled in the sheath; auricle absent; ligule very hairy, pointed	Rhizomes	5	6–8	4–6	20+	Excellent
Crabgrass *(Digitaria sanguanalis)*	Annual	Warm-season	Leaf rolled in the sheath; leaf blade flat, covered with short silky hairs on both surfaces; auricle small; ligule membranous, toothed, acute	Fibrous with stolons	2	2–4	2–4	1	Good
German foxtail millet *(Setaria italica)*	Annual	Warm-season	Foxtail-type panicle, large yellow bristles	Bunch	1	3–6	3–5	1	Fair
LEGUMES									
Alfalfa *(Medicago sativa)*	Perennial	Cool-season	Petioles: central leaflets larger than the other two leaflets; leaflets with ⅓ serrated margins and pointed tip; Flower: purple or blue	Deep taproot	2	2–4	4–5	4–6	Excellent
Alsike clover *(Trifolium hybridum)*	Perennial acts as biennial	Cool-season	Petioles: short petiolate, leaflets finely toothed, long stipule; Flower: white to pink	Taproot	–	–	–	2–4	Good
Birdsfoot trefoil *(Lotus corniculatus)*	Perennial	Cool-season	Leaflets: 5–3 leaflets at tip of petiole and 2 at base stipule; Flower: bright yellow	Taproot	5	2–4	2–4	3–5	Fair
Crownvetch *(Coronilla varia)*	Perennial	Cool-season	Petioles: narrow, pointed stipules; Leaf arrangement: pinnate with odd number of leaflets (no tendrils on leaf); Flower: pink to blue	Deep taproot and rhizomes	5	4–8	3–5	20+	Fair

(APPENDIX A. IDENTIFICATION OF MAJOR FORAGE SPECIES, cont.)

Common name/ scientific name [a]	Life cycle	Season of growth	ID key	Growth habit/ root structure	Ease of establishment (1-5) 1=easy	Harvest [b] Height (stubble) in.	Harvest [b] Frequency (weeks)	Life expectancy (years) [c]	Regrowth potential (fair, good, excellent)
Ladino clover *(Trifolium repens)*	Perennial	Cool-season	Same as common white clover but larger in size	Taproot and stolons	2	2–4	2–4	3–5	Good
Red clover *(Trifolium pratense)*	Biennial	Cool-season	Plant: entirely pubescent on both sides of leaves, leaflets usually with light-colored crescentic mark; Flower: two short-stalked leaves directly below flower; pink-purple	Taproot	2	2–4	2–4	2–3	Good
Sweet clover *(Melilotus alba; Melilotus officinalis)*	Biennial	Cool-season	Leaflets: all leaflets on short petioles, leaflets serrated for ⅔ of the margin of the leaf; Flower: yellow raceme	Long, deep taproot	3	2–4	3–5	2–3	Fair
White clover *(Trifolium repens)*	Perennial	Cool-season	Leaflets: heart-shaped with white crescent mark, very long petiole; Flower: white	Taproot and stolons	2	2–4	2–4	10+	Excellent
Crimson clover *(Trifolium incarnatum)*	Annual	Cool-season	Leaflets: slightly toothed, softly pubescent, stipule shallowly toothed and purplish-tipped; Flower: crimson head	Taproot	1	2–4	2–4	1	Fair
Hairy vetch *(Vicia villosa)*	Annual	Cool-season	Leaflets: pinnate, pointed, pubescent, terminal tendrils; Flower: blue-purple in one-sided raceme	Taproot	2	2–4	3–5	1	Fair
Lespedeza-Sericea *(Lespedeza cuneata)*	Perennial	Warm-season	Leaflets: pinnately trifoliate, small stipule; Flower: white or yellow	Taproot	4	2–4	2–4	10–20	Fair
Lespedeza-Korean *(Lespedeza stipulaceae)* **Common** *(Lespedeza striata)*	Annual	Warm-season	Leaflets: palmately trifoliate, leaflets connected at base, large stipule, noticeable parallel venation branching from midvein; Flower: purple-blue	Taproot	2	2–4	2–4	1	Fair
Pearl millet *(Pennisetum americanum)*	Annual	Warm-season	Compact cattail-type seedhead, hair under seedhead and around collar	Bunch	1	4–6	2–5	1	Fair

[a] *Names arranged by alphabetical order within life cycles. (Grasses: cool-season perennials, cool-season annuals, warm-season perennials, warm-season annuals. Same for legumes.)*

[b] *Shorter harvest stubble heights are recommended when grasses are grown in association with legumes. Harvest frequency partially depends on weather conditions and plant growth rate.*

[c] *Life expectancy of perennial grasses largely depends on the species' adaptation to the environment and management occurring on the site.*

APPENDIX B. COMMON AND SCIENTIFIC NAMES FOR COMMON NORTHEASTERN PASTURE PLANTS

Common name	Scientific name
Barley, little	*Hordeum pusillum*
Black medic	*Medicago lupulina*
Bluegrass, annual	*Poa annua*
Bouncing bet	*Saponaria officinalis*
Brome, downy	*Bromus tectorum*
Burdock, common	*Arctium minus*
Buttercups	*Ranunculus* spp.
Carrot, wild	*Daucus carota*
Cheat	*Bromus secalinus*
Cherry, black	*Prunus serotina*
Chickweed, common	*Stellaria media*
Chicory	*Cichorium intybus*
Clover species	*Trifolium* spp.
Cockle, white	*Lychnis alba*
Crabgrass, large	*Digitaria sanguinalis*
Dandelion	*Taraxacum officinale*
Deadnettle, red (purple)	*Lamium purpureum*
Dock species	*Rumex* spp.
Dock, curly	*Rumex crispus*
Fern, bracken	*Pteridium aquilinum*
Fleabane species	*Erigeron* spp.
Foxtail species	*Setaria* spp.
Foxtail, yellow	*Setaria lutescens*
Garlic/onion, wild	*Allium vineale canadense*
Groundcherry, smooth	*Physalis subglabrata*
Heal-all	*Prunella vulgaris*
Hemlock, poison	*Conium maculatum*
Hemp dogbane	*Apocynum cannabinum*
Henbit	*Lamium amplexicaule*
Horsenettle	*Solanum carolinense*
Horseweed (mare's tail)	*Conyza canadensis*
Ironweed, tall	*Vernonia altissima*
Jimsonweed	*Datura stramonium*
Johnsongrass	*Sorghum halepense*
Knapweed, spotted	*Centaurea maculosa*
Lambsquarters, common	*Chenopodium album*
Lettuce, prickly	*Lactuca serriola*

Common name	Scientific name
Locust, black	*Robinia pseudoacacia*
Loosestrife, purple	*Lythrum salicaria*
Mile-a-minute	*Polygonum perfoliatum*
Milkweed, common	*Asclepias syriaca*
Mullein, common	*Verbascum thapsus*
Mustard species	*Brassica, Thlaspi,* and *Lepidium* spp.
Mustard, garlic	*Alliaria petiolata*
Nettle, stinging	*Urtica dioica*
Nightshade species	*Solanum* spp.
Nightshade, eastern black	*Solanum ptycanthum*
Nutsedge, yellow	*Cyperus esculentus*
Panicum, fall	*Panicum dichotomiflorum*
Parsnip, wild	*Pastinaca sativa*
Pennycress, field	*Thlaspi arvense*
Pepperweed, field	*Lepidium campestre*
Pepperweed, Virginia	*Lepidium virginicum*
Pigweed species	*Amaranthus* spp.
Pokeweed, common	*Phytolacca americana*
Quackgrass	*Elytrigia repens*
Radish, wild	*Raphanus raphanistrum*
Ragweed, common	*Ambrosia artemisiifolia*
Red sorrel (Sheep sorrel)	*Rumex acetosella*
Redroot pigweed	*Amaranthus retroflexus*
Shepherd's-purse	*Capsella bursa-pastoris*
Smartweed, Pennsylvania	*Polygonum pensylvanicum*
Snakeroot, white	*Eupatorium rugosum*
St. Johnswort	*Hypericum perforatum*
Thistle, bull	*Cirsium vulgare*
Thistle, Canada	*Cirsium arvense*
Thistle, musk/plumeless	*Cardus nutans acanthoides*
Toadflax, yellow	*Linaria vulgaris*
Velvetleaf	*Abutilon theophrasti*
Wirestem muhly	*Muhlenbergia frondosa*
Yarrow, common	*Achillea millefolium*
Yellow rocket	*Barbarea vulgaris*

APPENDIX C. COMMON AND SCIENTIFIC NAMES FOR COMMON NORTHEASTERN PASTURE INSECTS

Common name	Scientific name
Beetle, toadflax	*Brachypterolus pulicarius*
Beetles, European	*Gallerucella calmariensis* and *G. pusilla*
Billbugs	*Sphenophorus* spp.
Buffalo fly	*Haematobia exigua*
Bush fly	*Musca vetustissima*
Clover root curculio	*Sitona hispidulus*
Eye worm	*Thelazia skrjabini* and *T. gulosa*
Face fly	*Musca autumnalis*
Field cricket	*Gryllus pensylvanicus*
Gray garden slug	*Deroceras reticulatum*
Ground crickets	*Allonemobius fasciatus* and *A. allardi*
Horn fly	*Haematobia irritans*
Housefly	*Musca domestica*
Japanese beetle	*Popillia japonica*
Marsh slug	*Deroceras laeve*
Parasitic rove beetle	*Aleochara tristis*
Potato leafhopper	*Empoasca fabae*
Root gall beetle	*Sphenoptera jugoslavica*
Rose seed chalcid	*Megastigmus aculeatus* var. *nigroflavus*
Rose stem girdler	*Agrilus aurichalceus*
Rosette weevil	*Trishosirocalus horridus*
Seedhead fly	*Chaetorellia acrolophi*
Seedhead gall flies	*Urophora affinis* and *U. quadrifasciata*
Seedhead weevil	*Bangasternums fausti*
Seedhead weevil	*Rhinocyllus conicus*
Weevil, clover	*Ischnopterapion virens*
Weevil, toadflax	*Gymnetron antirrhini*
Weevil, loosestrife	*Hylobius transversovittatus*
Wireworms	*Aeolus melillus* and *Melanotus* sp.

APPENDIX D. COMMON AND SCIENTIFIC NAMES FOR COMMON NORTHEASTERN PASTURE PATHOGENS

Common name	Scientific name
Bacterial wilt	*Corynebacterium insidiosum*
Fusarium root rot	*Fusarium* sp.
Fusarium wilt	*Fusarium oxysporum medicaginis*
Phytophthora root rot	*Phytophthora* sp.
Pythium root rot	*Pythium* sp.
Sclerotinia crown and stem rot	*Sclerotinia trifoliorium*
Verticillium wilt	*Verticillium albo-atrum*
White mold	*Sclerotinia sclerotiorum*

Abbreviations

ADF – acid detergent fiber
AUD – animal unit grazing days
AUM – animal unit month
CCE – calcium carbonate equivalent
CP – crude protein
cwt – hundred weight
DM – dry matter
IVDMD – in vitro dry-matter digestibility
LR – lime requirement
MiG – management-intensive grazing
NDF – neutral detergent fiber
ppm – parts per million
RYE – realistic yield expectation
SOC – soil organic carbon

Conversion Tables

Type of measurement	To convert:	Into:	Multiply by:
Length	centimeters (cm)	inches (in)	0.394
	feet (ft)	centimeters (cm)	30.48
	feet (ft)	inches (in)	12
	feet (ft)	yards (yd)	0.33
	inches (in)	feet (ft)	0.083
	inches (in)	millimeters (mm)	25.4
	inches (in)	centimeters (cm)	2.54
	meters (m)	inches (in)	39.37
	meters (m)	feet (ft)	3.281
	meters (m)	yards (yd)	1.094
	yards (yd)	feet (ft)	3
	yards (yd)	centimeters (cm)	91.44
	yards (yd)	meters (m)	0.9144
Area	acres	square feet (ft^2)	43,560
	acres	square yards (yd^2)	4,840
	acres	hectares (ha)	0.4047
	hectares (ha)	acres	2.471
	hectares (ha)	square meters (m^2)	10,000
	square inches (in^2)	square centimeters (cm^2)	6.452
	square centimeters (cm^2)	square inches (in^2)	0.155
	square feet (ft^2)	square centimeters (cm^2)	929.09
	square feet (ft^2)	square meters (m^2)	0.0929
	square meters (m^2)	square feet (ft^2)	10.76
	square meters (m^2)	square yards (yd^2)	1.196
Weight	grams (g)	ounces (oz)	0.0353
	kilograms (kg)	pounds (lb)	2.205
	metric tons (megagrams)	short tons	1.1023
	ounces (oz)	pounds (lb)	0.0625
	ounces (oz)	grams (g)	28.35
	pounds (lb)	ounces (oz)	16
	pounds (lb)	grams (g)	453.6
	short tons	metric tons (megagrams)	0.9078
Volume, solids	bushels (bu)	cubic feet (ft^3)	1.24
	bushels (bu)	cubic meters (m^3)	0.352
	bushels (bu)	liters (L)	35.24
	cubic feet (ft^3)	liters (L)	28.32
	cubic feet (ft^3)	U.S. gallons (gal)	7.48
	cubic feet (ft^3)	cubic inches (in^3)	1,728
	cubic feet (ft^3)	cubic yards (yd^3)	0.037
	cubic feet (ft^3)	bushels (bu)	0.804
	cubic inches (in^3)	milliliters (ml)	16.39
	cubic meters (m^3)	cubic yards (yd^3)	1.308
	cubic meters (m^3)	U.S. gallons (gal)	264.2
	cubic meters (m^3)	cubic feet (ft^3)	35.3
	cubic yards (yd^3)	cubic feet (ft^3)	27
	cubic yards (yd^3)	liters (L)	764.6

Conversion Tables *(cont.)*

Type of measurement	To convert:	Into:	Multiply by:
Volume, solids (cont.)	cubic yards (yd^3)	cubic meters (m^3)	0.765
	cubic yards (yd^3)	bushels (bu)	21.7
	gallons, U.S. dry (gal)	cubic inches (in^3)	269
	liters (L)	cubic inches (in^3)	61.02
	milliliters (mL)	cubic inches (in^3)	0.0610
	quarts, dry (qt)	cubic inches (in^3)	67.2
Volume, liquids	cubic centimeters (cm^3 or cc)	milliliters (mL)	1
	cups (c)	fluid ounces (fl oz)	8
	gallons, U.S. (gal)	cups (c)	16
	gallons, U.S. (gal)	cubic inches (in^3)	231
	gallons, U.S. (gal)	quarts (qt)	4
	gallons, U.S. (gal)	liters (L)	3.785
	gallons, U.S. (gal)	gallons, Imperial (gal)	0.833
	gallons, Imperial (gal)	cubic inches (in^3)	277.42
	gallons, Imperial (gal)	liters (L)	4.546
	gallons, Imperial (gal)	gallons, U.S. (gal)	1.20
	liters (L)	pints (pt)	2.113
	liters (L)	quarts (qt)	1.057
	liters (L)	gallons, U.S. (gal)	0.2642
	milliliters (mL)	fluid ounces (fl oz)	0.0338
	pints (pt)	fluid ounces (fl oz)	16
	pints (pt)	cups (c)	2
	pints (pt)	quarts (qt)	0.5
	pints (pt)	cubic inches (in^3)	28.87
	pints (pt)	liters (L)	0.4732
	fluid ounces (fl oz)	cubic inches (in^3)	1.805
	fluid ounces (fl oz)	tablespoons (Tbsp)	2
	fluid ounces (fl oz)	teaspoons (tsp)	6
	fluid ounces (fl oz)	milliliters (mL)	29.57
	quarts (qt)	fluid ounces (fl oz)	32
	quarts (qt)	cups (c)	4
	quarts (qt)	pints (pt)	2
	quarts (qt)	U.S. gallons, liquid (gal)	0.25
	quarts (qt)	cubic inches (in^3)	57.7
	quarts (qt)	liters (L)	0.9463
	tablespoons (Tbsp)	teaspoons (tsp)	3
	tablespoons (Tbsp)	milliliters (mL)	15
	teaspoons (tsp)	milliliters (mL)	5
Weight per volume	grams/cubic centimeter (g/cm^3)	pounds/cubic foot (lbs/ft^3)	62.3
	tablespoons/bushel (Tbsp/bu)	pounds/cubic yard (lbs/yd^3)	1 (approx.)
	pounds/cubic yard (lbs/yd^3)	ounces/cubic foot (oz/ft^3)	0.6
	ounces/cubic foot (oz/ft^3)	pounds/cubic yard (lbs/yd^3)	1.67
	pounds/cubic yard (lbs/yd^3)	grams/liter (g/L)	0.595
	kilograms/cubic meter (kg/m^3)	pounds/cubic yard (lbs/yd^3)	1.6821

Parts per million (ppm) conversions

- 1 milligram/liter = 1 ppm
- 1 ounce/gallon = 7,490 ppm
- 1 ounce/100 gallons = 75 ppm

percent fertilizer element x 75 = ppm of element in 100 gallons of water per ounce of fertilizer

For example, for a 9-45-15 fertilizer, the ppm nitrogen (N) in 100 gallons of water per ounce of fertilizer would be:
0.09 (percent N) x 75 = 6.75 ppm N in 100 gallons of water per ounce of 9-45-15

If you want 150 ppm N, and each ounce gives 6.75 ppm, then you need:
150 ÷ 6.75 = 22.22 ounces of 9-45-15 fertilizer in 100 gallons of water

Temperature conversion formulas

- To convert °C to °F: (°C x 9/5) + 32 = °F
- To convert °F to °C: (°F–32) x 5/9 = °C

Glossary

Acid detergent fiber – The residue remaining after boiling a forage sample in acid detergent solution. ADF contains cellulose, lignin, and ash, but not hemicellulose.

Acidic – Having a pH less than 7.

Acidifying effect – A characteristic of fertilizers causing the soil to become lower in pH (more acidic); measured in pounds of calcium carbonate required to neutralize the acidifying effect.

Alkaline – Having a pH greater than 7.

Allelopathic – Substances that are toxic to other plants, thereby suppressing their growth.

Amortization – The process of gradually reducing and finally eliminating a cost or debt over time.

Animal unit (AU) – 1,000 pounds of grazing animal(s).

Animal unit grazing day (AUD) – The amount of forage consumed by an animal unit in 24 hours.

Anthelmintic – A drug that kills parasitic worms.

Apical dominance – The hormonal control that upper buds exert on the growth of tiller buds in the plant.

Arthropods – Insects, spiders, and crustaceans.

Auricle – A claw- or hair-like part of a grass leaf, found on the front side of the leaf collar.

Axil – The point at which a leaf or branch is attached to the main stem.

Axillary bud – A bud growing in the axil.

Base saturation – The percentage of the cation exchange capacity of a soil that is occupied by basic cations (Ca^{++}, Mg^{++}, K^{+}). The remainder is occupied by the acidic cation (H^{+}).

Biennial – A plant that requires two growing seasons to complete its life cycle. Vegetative growth occurs in year one, followed by flower and seedhead formation in year two.

Biomass – The quantity of living matter, expressed as a concentration or weight per unit area; vegetative material.

Break crop – A crop that is planted to provide a change from an otherwise continuous crop.

Broadcast application – Spreading fertilizer or seeds over the soil surface without incorporation into the soil.

Bulk density – Mass of soil per unit volume.

C3 grasses – Temperate region or cool-season grasses such as orchardgrass that fix carbon during photosynthesis using two molecules of a 3-carbon acid.

C4 grasses – Tropical region or warm-season grasses such as bermudagrass that can actively transfer carbon dioxide to the site of photosynthesis using a 4-carbon acid.

Carbon sequestration – The binding of carbon within plant matter, underground, or in other sinks so that it is removed from the atmosphere.

Cation – A positively charged ion. Potassium (K^{+}) is an example.

Cation exchange capacity – The total amount of negative charges in the soil that can attract (adsorb) cations. It is expressed in terms of milliequivalents per 100 grams of soil at neutral pH (7) or at some other stated pH value.

Conventional seeding – Seeding done after conventional tillage with a moldboard plow and disk harrow. Seeds are placed into the soil at a depth appropriate for the species.

Creeping perennial – Perennial plant that spreads by vegetative structures such as rhizomes or stolons.

Cultipacker seeding – Seeding by means of a machine (a cultipacker) with two sets of corrugated rollers for trenching and packing soil around the seed and a seed-metering box or hopper that drops the seed.

Cultural practices – Nonchemical manipulations for enhancing the quality or yield of a crop.

Defoliation – The removal of leaves from a plant.

Denitrification – The conversion of nitrates in the soil to a gaseous form of nitrogen.

Drilled seeding – Seeding by means of a grain drill that has grass and legume seed attachments.

Endophyte – A fungus that lives within a plant.

Entry height – The plant height at which livestock enter a pasture for grazing.

Evapotranspiration – The loss of water from the soil by both evaporation and uptake followed by transpiration by the plant.

Exchangeable cations – Cations that are present on exchange sites on soil particles and are available to plants for uptake.

Exit height – The plant height at which livestock exit a pasture for grazing.

Frost seeding – Seeding done in late winter when alternate freezing and thawing of the soil surface is occurring. Seeds are broadcast, and the natural churning process covers the seed with soil.

Grass tetany – A nutritional condition in grazing ruminants in which the concentration of magnesium in the blood is too low for good health, resulting in paralysis and death of the animal.

Greenchop – Green manure. A cover crop that is turned back into the soil to add nutrients and organic matter.

Hardpan – An impervious soil layer that limits root penetration and restricts water movement.

Heavy metals – The trace elements found in relatively low concentrations in biosolids. These metals can be toxic to humans, animals, or plants at high concentrations. Includes arsenic, cadmium, copper, lead, mercury, molybdenum, nickel, selenium, and zinc.

Heavy soil – Soil that drains slowly and usually contains a lot of clay.

Hydrograph – Graph of variation of water flow over time.

Immobilization – The conversion of an element from the inorganic to the organic form in microbial or plant tissues.

In vitro dry-matter digestibility – A method of using rumen microbes in a controlled laboratory environment to digest forage samples to estimate their digestibility in the natural rumen of livestock.

Infiltration – Movement of one substance into another, as water into soil.

Inflorescence – The arrangement of flowers on a stem of a plant.

Inoculant – A preparation of beneficial microbes used to introduce those microbes to seeds, soils, or crops.

Ion – An element (as in a nutrient) with either a positive or negative charge.

Leaching – The movement and loss of dissolved nutrients as water percolates through soil.

Leader-follower grazing – A grazing system in which one herd of animals (leaders with a higher nutritional requirement) grazes a pasture lightly then moves on and a second group (the followers with a lower nutritional requirement) come in to finish grazing the pasture to the desired stubble height.

Leaf collar – The region of a grass leaf where the leaf blade meets the leaf sheath.

Leaf sheath – The lower part of a grass leaf that encircles or is folded around the stem of the seedhead.

Least significant difference – The minimum difference between two average values required for those values to be considered truly (statistically) different.

Light soil – Soil that drains quickly and usually contains a lot of sand.

Ligule – A tissue or hairy part of a grass leaf, found on the back side of the leaf collar and projecting from the top of the leaf sheath.

Lime requirement – The amount of pure calcium carbonate equivalent needed to raise the pH of a soil to a specified pH (often 6.5).

Liquid limit – The minimum mass of water content at which a sample of soil will barely flow under a standard treatment. Synonymous with "upper plastic limit."

Macroinvertebrates – Invertebrates large enough to be visible to the naked eye, such as insect larvae.

Management-intensive grazing – The managed grazing of pastures based on a depth of understanding of the biology and ecology of the plants and animals in the system with an emphasis on intensifying management of the system and not intensifying labor, capital inputs, or grazing pressure.

Meristem – The growing point of a plant that contains cells capable of dividing indefinitely and evolving into different types of tissue.

Microflora – The microbes present in a system, as in soil.

Micronutrients – Essential elements needed in very small amounts by plants. Examples include iron, zinc, and manganese.

Morphology – The physical characteristics of a plant that affect how the plant interacts with its environment and responds to management.

Nematode – A tiny wormlike organism that may feed on or in plants, including roots; they may be referred to as roundworms, threadworms, or eelworms.

Net primary productivity – Calculated as (plant photosynthesis – [respiration + factors reducing production]).

Neutral detergent fiber – The residue remaining after boiling a forage sample in neutral detergent solution. NDF represents the indigestible and slowly digestible components in plant cell walls (cellulose, hemicellulose, lignin, and ash). Used to estimate intake potential (inversely related).

Niche – Function of an organism within an ecological community.

Nitrogen fixation – The conversion of nitrogen gas to a plant-available source of nitrogen such as ammonia by chemical or biological means.

Nitrogen mineralization – The conversion of nitrogen from an organic form to an inorganic state via microbial decomposition.

Node – A place on a stem where a leaf is or was attached.

Nodulation – The formation of nodules on the roots of a leguminous plant by the infection of the roots with nitrogen-fixing *Rhizobium* bacteria of the proper strain for the legume species being infected.

Nonpoint source pollution – Water pollution that arises from an area such as a field and not a point source such as an industrial factory.

Organic matter – Plant and animal residue in the soil in various stages of decomposition.

Panicle – A loose, irregular compound flower cluster.

Percolation – The downward movement of water through soil.

Perennial – A plant that lives for more than 2 years and often reproduces by vegetative structures and seed production.

Petiole – The stem of a leaf.

pH – A measure of the acidity (pH less than 7) or alkalinity (pH greater than 7) of a solution; a pH of 7 is neutral; pH is known to influence nutrient availability.

Photoperiod – The length of the light versus dark part of the day; short days of winter compared to long days in summer that affect plant and animal growth and reproduction.

Preferential flow – Flow through macropores in soil, such as old root channels and earthworm burrows, which speed up water flow through soil and bypass sites where nutrients can be retained in soil.

Protozoa – A single-celled microorganism.

Prussic acid poisoning – Also known as hydrocyanic acid or HCN, prussic acid is a potentially lethal poison produced during digestion of plant species with high concentrations of cyanogenic glycosides. Species such as sorghum, sudangrass, and johnsongrass can accumulate cyanogenic glycosides particularly during drought and especially immediately after a drought has broken.

Pugging – The effect of livestock hooves on wet pasture soil; the hoofs leave depressions in the soil, and disrupt the sod in extreme cases.

Pure live seed – A term used to express the quality of seeds. It is expressed as a percentage of a seed lot that is pure seed that will germinate. The percentage is determined by multiplying the percentage of pure seed by the percentage of germination and dividing by 100.

Raceme – A simple inflorescence of flowers borne on a common elongated axis.

Renovation – The improvement of a pasture by the partial or complete destruction of the sod, plus liming, fertilizing, weed control, and seeding as may be required to establish or reestablish desired forage plants without an intervening crop.

Reproductive growth – Plant growth that results in the production of sexual reproductive tissue in flowers and seed heads.

Rhizome – An underground stem on plants such as smooth bromegrass that develops roots and aboveground tillers at nodes along its length.

Root pulsing – Cycling of root growth and death in response to periodic defoliation, as plants rebalance root and shoot functions.

Rumen-bypass – Protein in ruminant livestock feed that is not degraded in the rumen but passes intact into the lower intestinal tract, where it is digested and absorbed.

Rumen-degradable – Protein in ruminant livestock feed that is degraded and used in the rumen by the rumen microflora to make bacterial protein that passes into the lower intestinal tract, where it is digested and absorbed. In cases of excess degradable protein the excess leaves the rumen as ammonia and is excreted in the urine as urea, causing an energy expense to the animal for the conversion.

Runner – See stolon.

Runoff – Water leaving the land as surface discharge to a water course.

Salt index – The effect of a fertilizer having an ionic or drying effect similar to table salt in attracting water.

Selective translocated herbicides – Kill some plants and leave others unharmed. Translocated herbicides are absorbed through aboveground plant parts or roots and are then moved throughout the plant system to kill the entire plant. This is in contrast to contact herbicides, which kill only the parts of the plant to which the herbicide is applied.

Soil plasticity – The degree to which a soil sample can be deformed without rupture.

Soil strength – Measure of the capacity of a soil mass to withstand stresses without giving way to those stresses by becoming deformed.

Soil structure – The combination or arrangement of primary soil particles into secondary units, which are characterized on the basis of size, shape, and grade.

Stocking density – The number of animals present per unit land area at a given point in time.

Stolon – An aboveground runner on plants such as white clover; roots and develops leaves at nodes along its length.

Subsoil –The soil underneath topsoil.

Stratification – Self-sorting by plants to produce a mosaic of diverse plant communities at different locations.

Summer annual – An annual crop that grows well in the heat of summer and is sensitive to cool weather.

Symbiotic – The relationship of two organisms in a mutually beneficial often obligatory relationship.

Tap root – A primary descending root such as a carrot's root.

Tendril – Threadlike clinging part of a climbing plant.

Terminal bud – The bud at the terminal end of a shoot.

Texture (of soil) – The relative proportions of sand, silt, and clay particles in a mass of soil.

Tiller – A daughter plant arising from an axillary bud in a grass tiller that produces independent roots and stems.

Topsoil – The upper part of the soil, where most of the organic matter is found.

Translocated herbicide – An herbicide that is taken up by plants through the leaves and moves throughout the plant system, killing the entire plant, including the roots; also called systemic herbicide.

Transpiration – Evaporation of water from plant tissue to the atmosphere; transpiration occurs mainly through the stomata in the leaves.

Walk-in (or tread-in) seeding – A seeding done by broadcasting seed such as clover over a pasture then turning the livestock into the pasture, causing them to walk the seed into the soil as they graze; best accomplished in early spring or late summer.

Weed – A plant growing out of place.

Winter annual – An annual plant that is planted or arises from seed that germinates in the fall, overwinters, produces seed in the spring, and then dies.

References

CHAPTER 1

(1) Blaser, R. E. 1986. *Managing Forage for Animal Production*. Virginia Agricultural Experiment Station Bulletin 86–7.

(2) Davidson, R. L. 1978. Root systems—the forgotten components of pastures. In: J. R. Wilson (ed.). *Plant Relations in Pastures.* Hedges and Bell Pty. Ltd., Australia.

(3) Evans, L. T., Wardlaw, I. F., and Williams, C. N. 1964. In: C. Barnard (ed.). *Grass and Grasslands.* Macmillan, London.

(4) Ludlow, M. M. 1978. Light relations of pasture plants. In: J. R. Wilson (ed.). *Plant Relations in Pastures.* Hedges and Bell Pty. Ltd., Australia.

(5) Marten, G. 1985. Environmental and management limitations of legume based forage systems in the northern United States. In: *Forage Legumes for Energy-Efficient Animal Production.* Proc. Trilateral Workshop 1984, Palmerston North, N.Z. USDA, Washington, D.C.

(6) Nelson, C. J., and J. J. Volenec. 1995. Environmental and physiological aspects of forage management. In: R. F. Barnes, D. A. Miller, and D. J. Nelson (ed.). *Forages.* Iowa State University Press, Ames, IA.

(7) Rayburn, E. B., and S. B. Rayburn. 1998. A standardized plate meter for estimating pasture mass in on-farm research trials. *Agronomy Journal* 90: 238–241.

(8) Robson, A. D., and J. F. Loneragan. 1978. Responses of pasture plants to soil chemical factors other than nitrogen and phosphorus, with particular emphasis on the legume symbiosis. In: J. R. Wilson (ed.). *Plant Relations in Pastures.* Hedges and Bell Pty. Ltd., Australia.

(9) Smith, D. 1962. Carbohydrate root reserves in alfalfa, red clover, birdsfoot trefoil under several management schedules. *Crop Science* 2:75–78.

(10) Wilson, D. 1973. *Chemistry and Biochemistry of Herbages.* G. W. Butler and R. W. Bailey (ed.). Academic Press, London.

CHAPTER 2

(1) Carlassare, M., and H. D. Karsten. 2002. Species contribution to seasonal productivity of a mixed pasture under two sward grazing height regimes. *Agronomy Journal* 94: 840–850.

(2) Clark, E. A. 2001. Diversity and stability in humid temperate pastures. pp. 103–118, In: P. G. Tow and A. Lazenby (ed.). *Competition and Succession in Pastures.* CAB International Publishing, New York.

(3) Magurran, A. E. 1988. *Ecological Diversity and Its Measurement.* Princeton University Press: Princeton, NJ.

(4) Tracy, B. F., and M. A. Sanderson. 2003. Productivity and stability relationships in clipped pasture communities of varying species composition. *Crop Science* 44: 2180–2186.

CHAPTER 3

(1) Blaser, R. E., and E. L. Kimbrough. 1968. Potassium nutrition of forage crops with perennials. In: V. J. Kilmer, S. E. Younts, and N. C. Brady (ed.). *The Role of Potassium in Agriculture.* American Society of Agronomy, Madison, WI. 509 pp.

(2) Bryan, W. B., and K. C. Elliot. 1990. Effects of sample depth, and of lime and phosphorus applications on soil test levels in pasture soils. In: R. J. Write, V. C. Baligar, and R. P. Murrmann (ed.). *Plant-Soil Interactions at Low pH.* Kluwer Academic Publishers, Boston.

(3) Colyer, D., F. L. Alt, J. A. Balasko, P. R. Henderlong, G. A. Jung, and V. Thang. 1977. Economic optima and price sensitivity of N fertilization for six perennial grasses. *Agronomy Journal* 69: 514–517.

(4) Lathwell, D. J., and M. Peech. 1964. *Interpretation of Chemical Soil Tests.* Bull. 995. Cornell University Agricultural Experiment Station, Ithaca, NY. 40 pp.

(5) Matches, A. G. 1979. Management. In: R. C. Buckner and L. P. Bush (ed.). *Tall Fescue.* American Society of Agronomy, Madison, WI. 351 pp.

(6) Olson, R. A., K. D. Frank, P. H. Grabouski, and G. W. Rehm. 1981. Economic and agronomic impacts of varied philosophies of soil testing. *Agronomy Journal* 74: 492–499.

(7) Pearson, R. W., and C. S. Hoveland. 1974. Lime needs of forage crops. In: D. A. Mays (ed). *Forage Fertilization.* American Society of Agronomy, Madison, WI. 621 pp.

(8) Rayburn, E. B. 1987. *Pasture Management Facts and Figures for New York.* Seneca Trail RC&D, Franklinville, NY. 92 pp.

(9) Rayburn, E. B. 1996. *Forage Fertilization Based on Yield and Management Goals.* FS 5122. West Virginia University Extension Service, Morgantown, WV.

(10) Rayburn, E. B., M. H. Hall, W. Murphy, and L. Vough. 1998. Pasture production. In: C. R. Krueger and H. B. Pionke (ed.). *Pasture Management in the Northeast—Assessing Current Technologies, Research Directions, and Educational Needs.* NRAES–113. Northeast Regional Agricultural Engineering Service, Ithaca NY.

(11) Reid, W. S., W. L. Griffeth, R. Feuer, and R. B. Musgrave. 1967. *Effect of Lime Rate, Timing, and Placement on the Yields of Corn, Oats, and Hay.* Agronomy Mimeo. 67-13. Agronomy Department, Cornell University, Ithaca NY. 10 pp.

(12) Reinbott, T. M., and D. G. Blevins. 1994. Phosphorus and temperature effects on magnesium, calcium, and potassium in wheat and tall fescue leaves. *Agronomy Journal* 86: 523–529.

(13) Reinbott, T. M., and D. G. Blevins. 1997. Phosphorus and magnesium fertilization interaction with soil phosphorus level: Tall fescue yield and mineral element content. *Journal of Production Agriculture* 10: 260–265.

(14) Stout, W. L., and R. R. Schnabel. 1994. Soil drainage influence on biomass and nitrogen accumulation. *Agronomy Journal* 86: 111–116.

CHAPTER 4

(1) Barnes, R. F., D. A. Miller, and C. J. Nelson. 1995. *Forages, the Science of Grassland Agriculture,* 5th ed. Vols. 1 & 2. Iowa State University Press, Ames, IA.

(2) Barrow, N. J. 1967. Some aspects of the effects of grazing on the nutrition of pastures. *Journal of Australian Agricultural Science* 33: 254–262.

(3) Betteridge, K., W. G. K. Andrewes, and J. R. Sedcole. 1986. Intake and excretion of nitrogen, potassium, and phosphorus by grazing steers. *Journal of Agricultural Science* (Camb.) 106: 393–404.

(4) Blackshaw, J. K., and A. W. Blackshaw. 1994. Heat stress in cattle and the effect of shade on production and behaviour: a review. *Australian Journal of Experimental Agriculture* 34: 285–295.

(5) Curll, M. L., and R. J. Wilkins. 1983. The comparative effects of defoliation, treading, and excreta on a *Lolium perenne-Trifolium repens* pasture grazed by sheep. *Journal of Agricultural Science* 100: 451–460.

(5a) Dairy One Forage Testing Lab. 1994 annual summary of forage and grain analysis. Ithaca, N.Y.

(6) Dalrymple, R. L., R. Stevens, T. Carroll, and B. Flatt. 1994. Forage production benefits from nutrient recycling via beef cattle and how to manage for nutrient recycling in a grazing cell. pp. 269–273. In: *American Forage and Grassland Council Proceedings,* Lancaster, PA. 6–10 March 1994. American Forage and Grassland Council, Georgetown, TX.

(7) During, C., and W. C. Weeda. 1973. Some effects of cattle dung on soil properties, pasture production, and nutrient uptake. *New Zealand Journal of Agricultural Research* 16: 423–430.

(8) Edwards, D. R. 1996. Recycling livestock wastes on pastures. pp. 45–63. In: R. E. Joost and C. A. Roberts (ed.). *Nutrient Cycling in Forage Systems.* Columbia, MO. 7–8 March 1996. Potash and Phosphate Institute, Manhattan, KS.

(9) Evanylo, G. K. 1999. *Land Application of Biosolids for Agricultural Purposes in Virginia.* Virginia Cooperative Extension Publ. No. 452–300.

(10) Evanylo, G., and P. Peterson. 2000. *Availability of N in Biosolids for Tall Grass Hay Production.* Final report to the T. M. Hepler Endowment Committee, Virginia Tech. 9 pp.

(11) Field, T. R. O., P. R. Ball, and P. W. Theobald. 1985. Leaching of nitrate from sheep-grazed pastures. *Proceedings of New Zealand Grassland Association* 46: 209–214.

(12) Fraser, P. M., K. C. Cameron, and R. R. Sherlock. 1994. Lysimeter study of the fate of nitrogen in animal urine returns to irrigated pasture. *European Journal of Soil Science* 45: 439–447.

(13) Gerrish, J. R., J. R. Brown, and P. R. Peterson. 1993. Impact of grazing cattle on distribution of soil minerals. pp. 66–70. In: *American Forage and Grassland Council Proceedings.* Des Moines, IA. 29–31 March 1993. American Forage and Grassland Council, Georgetown, TX.

(14) Haynes, R. J., and P. H. Williams. 1993. Nutrient cycling and soil fertility in the grazed pasture ecosystem. *Advances in Agronomy* 49.

(15) Hilder, E. J. 1966. Distribution of excreta by sheep at pasture. pp. 977–981. *Proceedings Xth International Grassland Congress.* Helsinki, Finland. 7–16 July 1966. International Grassland Council, Helsinki, Finland.

(16) Jarvis, S. C., and D. J. Hatch. 1994. Potential for denitrification at depth below long-term grass swards. *Soil Biology and Biochemistry* 26: 1629–1636.

(17) Lemunyon, J., and J. B. Cropper. 1992. Role of plants in waste management. Ch. 6. In: *Agricultural Waste Management Field Handbook.* USDA-Soil Conservation Service. Washington, D.C.

(18) Lory, J., and R. Kallenbach. 1999. Soil fertility management and nutrient cycling. pp. 73–80. In: J. Gerrish and C. Roberts (ed.). *Missouri Grazing Manual.* MU-Extension, Universisty of Missouri.

(19) Miner, J. R., J. C. Buckhouse, and J. A. Moore. 1992. Evaluation of off-stream water source to reduce impact of winter fed range cattle on stream water quality. *Rangelands* 14: 65–75.

(20) Morrison. 1956. *Feeds and Feeding.* p. 640. Morrison Publishing Co., Ithaca, NY.

(21) Morton, J. D., and D. B. Baird. 1990. Spatial distribution of dung patches under sheep grazing. *New Zealand Journal of Agricultural Research* 33: 285–294.

(22) National Research Council (NRC). 1988. *Nutrient Requirements of Dairy Cattle.* p. 99–100. National Academy Press, Washington, D.C.

(23) NRC. 1996. *Nutrient Requirements of Beef Cattle.* p. 23. National Academy Press, Washington, D.C.

(24) O'Connor, K. F. 1974. Nitrogen in agrobiosystems and its environmental significance. *New Zealand Journal of Agricultural Science* 8: 137–148.

(25) Owens, L. B., W. M. Edwards, and R. W. Keuren. 1983. Surface runoff water quality comparisons between unimproved pasture and woodland. *Journal of Environmental Quality* 12: 518–522.

(26) Petersen, R. G., H. L. Lucas, and W. W. Woodhouse, Jr. 1956. The distribution of excreta of freely grazing cattle and its effect on pasture fertility. I. Excretal distribution. *Agronomy Journal* 48: 440–444.

(27) Peterson, P. R., and J. R. Gerrish. 1995. Grazing management affects manure distribution by beef cattle. pp. 170–174. In: *American Forage and Grassland Council Proceedings.* American Forage and Grassland Council, Georgetown, TX.

(28) Peterson, P. R., and J. R. Gerrish. 1996. Grazing systems and spatial distribution of nutrients in pastures: Livestock management considerations. pp. 203–212. In: R. E. Joost and C. A. Roberts (ed.). *Nutrient Cycling in Forage Systems.* Potash and Phosphate Institute, Manhattan, KS.

(29) Reid et al. 1955. *Journal of Dairy Science* 38:1344.

(30) Richards, I. R., and K. M. Wolton. 1976. The spatial distribution of excreta under intensive cattle grazing. *Journal of the British Grassland Society* 31: 89–92.

(31) Russelle, M. P. 1996. Nitrogen cycling in pasture systems. pp. 125–166. In: R. E. Joost and C. A. Roberts (ed.). *Nutrient Cycling in Forage Systems.* Potash and Phosphate Institute, Manhattan, KS.

(32) Scholefield, D., D. R. Lockyer, D. C. Whitehead, and K. C. Tyson. 1991. A model to predict transformations and losses of nitrogen in UK pastures grazed by beef cattle. *Plant and Soil* 132: 165–177.

(33) Schomberg, H. H., J. A. Stuedemann, A. J. Franzluebbers, and S. R. Wilkinson. 2000. Spatial distribution of extractable phosphorus, potassium, and magnesium as influenced by fertilizer and tall fescue endophyte status. *Agronomy Journal* 92: 981–986.

(34) Soil Conservation Service. 1992. *Agricultural Waste Management Field Handbook.* U.S. Dept. Agric., Washington, D.C.

(35) Sommers, L. E. 1977. Chemical composition of sewage sludges and analysis of their potential use as fertilizers. *Journal of Environmental Quality* 6: 225–239.

(36) Steele, K. W, M. J. Judd, and P. W. Shannon. 1984. Leaching of nitrate and other nutrients from a grazed pasture. *New Zealand Journal of Agricultural Research* 27: 5–11.

(37) Stout, W. L., W. J. Gburek, R. R. Schnabel, G. J. Folmar, and S. R. Weaver. 1998. Soil-climate effects on nitrate leaching from cattle excreta. *Journal of Environmental Quality* 27: 992–998.

(38) Svensson, L. 1994. Ammonia volatilization following application of livestock manure to arable land. *Journal of Agricultural Engineering Research* 58: 241–260.

(39) Virginia Department of Conservation and Recreation. 1995. *Virginia Nutrient Management Standards and Criteria.* Richmond, VA. 64 pp.

(40) West, C. P., A. P. Mallarino, W. F. Wedin, and D. B. Marx. 1989. Spatial variability of soil chemical properties in grazed pastures. *Soil Science Society of America Journal* 53: 784–789.

(41) White, S. L., R. E. Sheffield, S. P. Washburn, L. D. King, and J. T. Green, Jr. 2001. Spatial and time distribution of dairy cattle excreta in an intensive pasture system. *Journal of Environmental Quality* 30: 2180–2187.

(42) Whitehead, D. C. 1970. *The Role of Nitrogen in Grassland Productivity.* Bulletin No. 48. Commonwealth Agric. Bureau, Farnham Royal, Bucks, UK.

(43) Wilkinson, S. R., and R. W. Lowrey. 1973. Cycling of mineral nutrients in pasture ecosystems. pp. 247–315. In: G. W. Butler and R. W. Bailey (ed.). *Chemistry and Biochemistry of Herbage,* Vol. 2. Academic Press, New York, NY.

(44) Wilkinson, S. R., J. A. Stuedemann, and D. P. Belesky. 1989. Soil potassium distribution in grazed K-31 tall fescue pastures as affected by fertilization and endophytic fungus infection level. *Agronomy Journal* 81: 508–512.

(45) Williams, P. H., and R. J. Haynes. 1992. Balance sheet of phosphorus, sulphur, and potassium in a long-term grazed pasture supplied with superphosphate. *Fertility Research* 31: 51–60.

CHAPTER 5

(1) Aiello, S. E. 1998. *The Merck Veterinary Manual,* 8th ed. Merck and Co, Inc., Whitehouse Station, NJ. 2305 pp.

(2) Alderfer, R. B., and R. R. Robinson. 1947. Runoff from pastures in relation to grazing intensity and soil compaction. *Agronomy Journal* 39: 948–958.

(3) Ball, R., D. R. Keeney, P. W. Theobald, and P. Nes. 1979. Nitrogen balance in urine-affected areas of a New Zealand pasture. *Agronomy Journal* 71: 309–314.

(4) Ball, P. R., and J. C. Ryden. 1984. Nitrogen relationships in intensively managed temperate grasslands. *Plant and Soil* 76: 23–33.

(5) Barbour, M. T., J. Gerritsen, B. D. Snyder, and J. B. Stribling. 1999. *Rapid Bioassessment Protocols for Use in Streams and Wadeable Rivers: Periphyton, Benthic Macroinvertebrates and Fish,* 2nd ed. EPA 841-B-99-002. U.S. Environmental Protection Agency, Office of Water, Washington, D.C.

(6) Baxter-Potter, W. R., and M. W. Gilliland. 1988. Bacterial pollution in runoff from agricultural lands. *Journal of Environmental Quality* 17: 27–34.

(7) Breeuwsma, A., J. G. A. Reijerink, and O. F. Schoumans. 1995. Impact of manure on accumulation and leaching of phosphate in areas on intensive livestock farming. pp. 239–249, In: K. Steele (ed.). *Animal Waste and the Land-Water Interface.* Lewis Publishers, New York, NY.

(8) Chaney, E., W. Elmore, and W. S. Platts. 1990. *Livestock Grazing on Western Riparian Areas.* U.S. Environmental Protection Agency, Northwest Resource Information Center, Inc. Eagle, ID. July. 45 pp.

(9) Chardon, W. J., O. Oenema, P. del Castilho, R. Vriesema, J. Japenga, and D. Blaauw. 1997. Organic phosphorus in solutions and leachates from soils treated with animal manure. *Journal of Environmental Quality* 26: 372–378.

(10) Clark, E. A. *Cattle and Creeks—Can They Coexist?* University of Guelph, Guelph, Ontario. February, 1998. 8 pp.

(11) Clary, W. P., and B. F. Webster. 1989. *Managing Grazing of Riparian Areas in the Intermountain Region.* USDA, Forest Service. Intermountain Research Station. General Technical Report INT-263. May. 11 pp.

(12) Colman, E. A. 1953. *Vegetation and Watershed Management.* The Ronald Press Company, New York, NY. pp. 120, 178.

(13) Corps of Engineers, New England Division, U.S. Army. 1991. *Buffer Strips for Riparian Zone Management.* January. 56 pp.

(14) Cralle, H. T., and G. H. Heichel. 1981. Nitrogen fixation and vegetative regrowth of alfalfa and birdsfoot trefoil after successive harvests or floral debudding. *Plant Physiology* 67: 898–905.

(15) Cuttle, S. P., and D. Scholfield. 1994. Management options to limit nitrate leaching from grassland. pp. 138–150, In: *Transactions of the 15th World Congress of Soil Science.* International Soil Science Society. Acapulco, Mexico.

(16) Donelly, C. W. 1994. *Listeria monocytogenes.* In: Y. H. Hui, J. R. Gorham, K. D. Murrell, and D. O. Cliver (ed.). *Foodborne Disease Handbook: Diseases Caused by Bacteria,* vol. 1. Marcel Dekker, Inc., New York, NY. pp. 215–252.

(17) Dormaar, J. F., and S. Smoliak. 1985. Recovery of vegetative cover and soil organic matter during revegetation of abandoned farmland in a semi-arid climate. *Journal of Range Management* 51: 487–491.

(18) Duda, A. M., and D. S. Finan. 1983. Influence of livestock on nonpoint source nutrient levels of streams. *Transactions, American Society of Agricultural Engineering* 26: 1710–1716.

(19) Dyer, M. I., C. L. Turner, and T. R. Seastedt. 1998. Biotic interactivity between grazers and plants: Relationships contributing to atmospheric boundary layer dynamics. *Journal of Atmospheric Science* 55: 1247–1259.

(20) Eriksen, J. F. P. Vinther, and K. Søergaard. 2004. Nitrate leaching and N_2-fixation in grasslands of different composition, age, and management. *Journal of Agricultural Science* 142: 141–151.

(21) Fayer, R., J. M. Trout, and M. C. Jenkins. 1998. Infectivity of *Cryptosporidium parvum* oocysts stored in water at environmental temperatures. *Journal of Parasitology* 84: 1105–1108.

(22) Franzluebbers, A. J., F. M. Hons, and D. A. Zuberer. 1998. In situ and potential CO_2 evolution from a Fluventic Ustochrept in southcentral Texas as affected by tillage and cropping intensity. *Soil Tillage Research* 47: 303–308.

(23) Garber, L. P., M. D. Salman, H. S. Hurd, T. Keefe, and J. L. Schlater. 1994. Potential risk factors for *Cryptosporidium* infection in dairy calves. *Journal of the American Veterinary Medicine Association* 205: 86–91.

(24) Garwood, E. A., and J. C. Ryden. 1986. Nitrate loss through leaching and surface runoff from grassland: effects of water supply, soil type and management. pp. 90–113, In: H. G. van der Meer, J. C. Ryden, and G. C. Ennik (ed.) *Nitrogen Fluxes in Intensive Grassland Systems.* Martinus Nijoff Publishers, Dordrecht, The Netherlands.

(25) Gerrish, J. R., P. R. Peterson, and J. R. Brown. 1995. Grazing management affects soil phosphorus and potassium levels. pp. 174–179, In: *American Forage and Grassland Council Proceedings,* March 1995, Lexington, KY.

(26) Gradwell, M.W. 1965. Soil moisture deficiencies in puddled pastures. *New Zealand Journal of Agricultural Research* 9: 127–136.

(27) Greenwood, K. L., D. A. MacLeod, and K. J. Hutchinson. 1997. Long-term stocking rate effects on soil physical properties. *Australian Journal of Experimental Agriculture* 37: 413–419.

(28) Harper, L. A., V. R. Catchpoole, R. Davis, and K. L. Weir. 1983. Ammonia volatilization: Soil, plant and microclimate effects on diurnal and seasonal fluctuations. *Agronomy Journal* 75: 212–218.

(29) Harper, L. A., V. R. Catchpoole, and I. Vallis. 1983. Gaseous ammonia transport in a cattle-pasture system. p. 353–372, In: R. Lowrance et al. (ed.). *Nutrient Cycling in Agricultural Ecosystems,* University of Georgia, College of Agriculture Experiment Station. Special Publication 23.

(30) Hassink, J. 1994. Effects of soil texture and grassland management on soil organic C and N and rates of C and N mineralization. *Soil Biology and Biochemistry* 26: 1221–1231.

(31) Hassink, J., and J. J. Neetson. 1991. Effects of grassland management on the amounts of soil organic N and C. *Netherlands Journal of Agricultural Science* 39: 225–236.

(32) Haygarth, P. M., and S. C. Jarvis. 1996. Pathways and forms of phosphorus losses from grazed grassland hillslopes. pp. 283–294, In: M. G. Anderson and S. M. Brooks (ed.). *Advances in Hillslope Processes.* John Wiley and Sons, Chichester, UK.

(33) Haynes, R. J., and P. H. Williams. 1993. Nutrient cycling and soil fertility in the grazed pasture ecosystem. pp. 119–199, In: D. L. Sparks (ed.). *Advances in Agronomy,* vol. 49. Academic Press, Inc., New York, NY.

(34) Hoglund, J. H. 1985. Grazing intensity and soil nitrogen accumulation. *Proceedings of the New Zealand Grassland Association* 46: 65–69.

(35) Jarvis, S. C., D. J. Hatch, and D. H. Roberts. 1989. The effects of grassland management on nitrogen losses from grazed swards through ammonia volatilization, the relationship to excretal N returns from cattle. *Journal of Agricultural Science* (Cambridge) 112: 205–216.

(36) Khaleel, R., K. R. Reddy, and M. R. Overcash. 1980. Transport of potential pollutants in runoff water from land areas receiving animal wastes: A review. *Water Research* 14: 421–436.

(37) Kleinman, P. J. A., W. J. Gburek, A. N. Sharpley, W. L. Stout, and R. B. Bryant. 2002. Impact of pastured dairy cattle on stream and near-stream areas on a Catskill farm, New York. pp. 30–38, In: *Evaluation and Management of Phosphorus in the Town Brook Watershed: Initial Best Management Practice Investigations.* Final Report for Contract No. C004171. New York Department of Environmental Protection, Shokan, NY.

(38) Kok, H., R. K. Taylor, R. E. Lamond, and S. Kessen. 1996. *Soil Compaction, Problems and Solutions.* Kansas State University Cooperative Extension Service, Manhattan, KS. Crops and Soils Bulletin AF-115.

(39) Krenzer, E. G., Jr., C. F. Chee, and J. F. Stone. 1989. Effects of animal traffic on soil compaction in wheat pastures. *Journal of Production Agriculture* 2: 246–249.

(40) Langer, R. H. M. 1994. *Pastures: Their Ecology and Management.* Oxford University Press, Auckland, New Zealand. pp. 181–187.

(41) Larson, L. L., and S. L. Larson. 1996. Riparian shade and stream temperature: A perspective. *Rangelands* 18: 149–152.

(42) Lovell, A. D., S. R. Wilkinson, J. A. Stuedemann, D. H. Seman, and A. J. Franzluebbers. 1997. Broiler litter and grazing pressure impacts on soil organic C and N pools. *Agronomy Abstracts,* p. 217. American Society of Agronomy, Madison, WI.

(43) Lowrance, R., L. Altier, D. Newbold, R. Schnabel, P. Groffman, J. Denver, D. Correll, J. W. Gilliam, J. Robinson, R. Brinsfield, K. Stave, W. Lucas, and A. Todd. 1994. *Riparian Forest Buffer Systems in the Chesapeake Bay Watersheds—Potential Uses.* Draft. USDA/ARS Southeast Watershed Research Laboratory, Tifton, GA. 110 pp.

(44) McDonald, A., D. Kay, and A. Jenkins. 1982. Generation of fecal and total coliform surges by stream flow manipulation in the absence of normal hydrometeorological stimuli. *Applied Environmental Microbiology* 44(2): 292–300.

(45) McFarland, A., and L. Hauck. 1995. *Livestock and the Environment: Scientific Underpinnings for Policy Analysis.* Report No. 1, Texas Institute for Applied Environmental Research, Stephenville, TX. Tarleton State University, Stephenville, TX. 140 pp.

(46) Milchunas, D. G., and W. K. Lauenroth. 1993. Quantitative effects of grazing on vegetation and soils over a global range of environments. *Ecological Monographs* 63: 327–366.

(47) Miller, T. 1997. Can trout and cows coexist? *Natural Resources Report.* University of Wisconsin-Madison, WI. pp. 4–5.

(48) Miner, J. R., J. C. Buckhouse, and J. A. Moore. 1992. Evaluation of off-stream water source to reduce impact of winter-fed range cattle on stream water quality. *Rangelands* 14: 65–75.

(49) Nelson, P. N., E. Cotsaris, and J. M. Oades. 1996. Nitrogen, phosphorus, and organic carbon in streams draining two grazed catchments. *Journal of Environmental Quality* 25: 1221–1229.

(50) New York City Department of Environmental Protection. 1995. *Evaluation and Prioritization of the Overall Phosphorus Reduction Strategy in the Catskill and Delaware Watersheds.* New York Department of Environmental Protection, Shokan, NY.

(51) Nguyen, M. L., and K. M. Goh. 1992. Status and distribution of soil sulphur fractions, total nitrogen and organic carbon in camp and non-camp of grazed pastures supplied with long-term superphosphate. *Biology and Fertility of Soils* 14: 181–190.

(52) O'Connor, K. 1957. Influences of wheel and foot treading on soils under grasslands. *Soil Science Society of America Proceedings* 2: 35–37.

(53) Olness, A. E., S. J. Smith, E. D. Rhoades, and R. G. Menzel. 1975. Nutrient and sediment discharge from agricultural watersheds in Oklahoma. *Journal of Environmental Quality* 4: 331–336.

(54) Owens, L. B., W. M. Edwards, and R. W. Van Keuren. 1997. Runoff and sediment losses resulting from winter feeding on pastures. *Journal of Soil and Water Conservation* 52: 194–197.

(55) Owens, L. B., R. W. Van Keuren, and W. M. Edwards. 1982. Environmental effects of a medium-fertility 12-month pasture program: I. Hydrology and soil loss. *Journal of Environmental Quality* 11: 236–240.

(56) Paine, L. 1998. Impacts of intensive rotational grazing on stream ecology and water quality. *Great Lakes Basin Grazier* 1: 1–7.

(57) Palone, R. 1990. The benefits of buffer areas for moderating stream water temperature. *Forest Management Update,* Issue 11. USDA, Forest Service, Morgantown, WV. July, pp. 14–15.

(58) Patni, N. K., H. R. Toxopeus, and P. Y. Jui. 1985. Bacterial quality of runoff from manured and non-manured cropland. *Transactions of the American Society of Agricultural Engineers* 28(6): 1871–1877.

(59) Paustian, K., G. P. Robertson, and E. T. Elliot. 1995. Management impacts on carbon storage and gas fluxes (CO_2, CH_4) in mid-latitude cropland. pp. 69–83, In: R. Lal, J. Kimble, E. Levine, and B. A. Stewart (ed.). *Soil Management and Greenhouse Effect.* Advances in Soil Science, Lewis Publishers, Boca Raton, FL.

(60) Petersen, R. C. 1992. The RCE: a riparian, channel, and environmental inventory for small streams in the agricultural landscape. *Freshwater Biology* 27: 295–306.

(61) Petersen, R. G., H. L. Lucas, and W. W. Woodhouse, Jr. 1956. The distribution of excreta by freely grazing cattle and its effect on pasture fertility: I. Excretal distribution. *Agronomy Journal* 48: 440–444.

(62) Petersen, R. G., W. W. Woodhouse, Jr., and H. L. Lucas. 1956. The distribution of excreta by freely grazing cattle and its effect on pasture fertility: I. The effect of returned excreta on the residual concentration of some fertilizer elements. *Agronomy Journal* 48: 444–449.

(63) Peterson, P. R., and J. R. Gerrish. 1996. Grazing systems and spatial distribution of nutrients in pastures: Livestock management considerations. pp. 203–212, In: R. E. Joost and C. A. Roberts (ed.). *Nutrient Cycling in Forage Systems.* Potash Phosphate Institute and Foundation for Agronomic Research, Manhattan, KS.

(64) Reeder, J. D., G. E. Schuman, J. A. Morgan, D. R. Lecain, and R. H. Hart. 1998. Impact of livestock grazing on the carbon and nitrogen balance of a shortgrass steppe. *Agronomy Abstracts,* p. 291. Am. Soc. Agron., Madison, WI.

(65) Russell, J. S. 1986. Improved pastures. pp. 374–396, In: J. S. Russell and R. F. Isbell (ed.) *Australian Soils: The Human Impact.* University of Queensland Press, St. Lucia, Queensland, Australia.

(66) Ryden, J. C., P. R. Ball, and E. A. Garwood. 1984. Nitrate leaching from grassland. *Nature* (London) 311: 50–53.

(67) Schnabel, R. R., A. J. Franzlubbers, W. L. Stout, M. A. Sanderson, and J. A. Steudemann. 2000. The effects of pasture management practices. pp. 291–322, In: R. F. Follett, J. M. Kimble, and R. Lal (ed.). *The Potential of U.S. Grazing Lands to Sequester Carbon and Mitigate the Greenhouse Effect.* Lewis Publishers, Boca Raton, FL.

(68) Scott, C. A., M. F. Walter, E. S. Brooks, J. Boll, M. B. Hes, and M. D. Merrill. 1998. Impacts of historical changes in land use and dairy herds in water quality in the Catskills Mountains. *Journal of Environmental Quality* 27: 1410–1417.

(69) Sharpley, A. N., and S. Rekolainen. 1997. Phosphorus in agriculture and its environmental implications. pp. 1–54, In: H. Tunney, O. T. Carton, P. C. Brookes, and A. E. Johnston (ed.). *Phosphorus Loss from Soil to Water.* CAB International Press, Cambridge, England.

(70) Sharpley, A. N., and J. K. Syers. 1976. Phosphorus transport in surface run-off as influenced by fertiliser and grazing cattle. *New Zealand Journal of Science* 19: 277–282.

(71) Sharpley, A. N., and J. K. Syers. 1979. Loss of nitrogen and phosphorus in tile drainage as influenced by urea application and grazing animals. *New Zealand Journal of Agricultural Research* 22: 127–131.

(72) Sharpley, A. N., J. J. Meisinger, A. Breeuwsma, T. Sims, T. C. Daniel, and J. S. Schepers. 1997. Impacts of animal manure management on ground and surface water quality. p. 173–242, In: J. Hatfield (ed.). *Effective Management of Animal Waste as a Soil Resource.* Lewis Publishers, Boca Raton, FL.

(73) Shere, J. A., K. J. Bartlett, and C. W. Kaspar. 1998. Longitudinal study of *Escherichia coli* O157:H7 dissemination on four dairy farms in Wisconsin. *Applied Environmental Microbiology* 64: 1390–1399.

(74) Snaydon, R. W. 1981. The ecology of grazed pastures. pp. 13–31, In: F. H. W. Morley (ed.). *Grazing Animals.* Elsevier, Amsterdam, The Netherlands.

(75) Spedding, C. R. W. 1971. *Grassland Ecology.* Oxford University Press at the Clarendon Press, UK.

(76) Steele, K. W., and I. Vallis. 1987. The nitrogen cycle in pastures. pp. 274–290, In: J. R. Wilson (ed.). *Advances in Nitrogen Cycling in Agricultural Ecosystems. Proceedings of the Symposium on Advances in Nitrogen Cycling in Agricultural Ecosystems.* Brisbane, Australia. C.A.B. Int., England.

(77) Steenvoorden, J., H. Fonck, and H. P. Oosterom. 1986. Losses of nitrogen from intensive grassland systems by leaching and surface runoff. pp. 85–97, In: H. G. van der Meer, J. C. Ryden, and G. C. Ennik (ed.). *Nitrogen Fluxes in Intensive Grassland Systems.* Martinus Nijoff Publishers, Dordrecht, The Netherlands.

(78) Stout, W. L., S. L. Fales, L. D. Muller, R. R. Schnabel, G. F. Elwinger, and S. R. Weaver. 2000. Assessing the effect of management intensive grazing on water quality in the northeast U.S. *Journal of Soil and Water Conservation* 55: 238–243.

(79) Stout, W. L., S. L. Fales, L. D. Muller, R. R. Schnabel, W. E. Priddy, and G. R. Elwinger. 1997. Nitrate leaching from cattle urine and feces in northeast USA. *Soil Science Society of America Journal* 61: 1787–1794.

(80) Stuedemann, J. A., A. J. Franzluebbers, D. H. Seman, S. R. Wilkinson, R. R. Bruce, A. D. Lovell, and S. W. Knapp. 1998. Role of the grazing animal in soil carbon restoration in the Southern Piedmont. *Agronomy Abstracts,* p. 57. American Society of Agronomy, Madison, WI.

(81) Tainton, N. M., C. D. Morris, and M. B. Hardy. 1996. Complexity and stability in grazing systems. In: J. Hodgson and A. W. Illius (ed.). *The Ecology and Management of Grazing Systems.* CAB International, Wallingford, England.

(82) Till, A. R., and A. P. Kennedy. 1981. The distribution in soil and plant of 35S sulfur isotope from sheep excreta. *Australian Journal of Agricultural Research* 32: 339–351.

(83) Tothill, J. C. 1978. Comparative aspects of the ecology of pastures. In: J. R. Wilson (ed.). *Plant Relations in Pastures.* CSIRO, Australia, pp. 385–420.

(84) Vallentine, J. F. 1990. *Grazing Management.* Academic Press, Inc., San Diego, CA. pp. 50, 101–104.

(85) Van den Pol-Van Dasselaar, A., and E. A. Lantinga. 1995. Modelling the carbon cycle of grassland in the Netherlands under various management strategies and environmental conditions. *Netherlands Journal of Agricultural Science* 43: 183–194.

(86) Walker, T. W., B. K. Tharpa, and A. F. R. Adams. 1959. Studies on soil organic matter: 3. Accumulation of C, N, S, organic and total P in improved grassland soils. *Soil Science* 87: 135–140.

(87) Watkin, B. R., and R. J. Clements. 1978. The effects of grazing animals on pastures. In: J. R. Wilson (ed.). *Plant Relations in Pastures.* Commonwealth Scientific and Industrial Research Org., East Melbourne, Australia. pp. 273–289.

(88) Welsh, D. J. 1992. *Riparian Forest Buffers.* USDA, Forest Service, Forest Resources Management, Northeastern Area, Radnor, PA. NA-PR-07-91.

(89) Wesley, I. V., S. J. Wells, K. M. Harmon, A. Green, L. Schroeder-Tucker, M. Glover, and I. Sudduque. 2000. Fecal shedding of *Campylobacter* and *Arcobacter* spp. in dairy cattle. *Applied Environmental Microbiology* 66(5): 1994–2000.

(90) West, C. P., A. P. Mallarino, W. F. Wedin, and D. B. Marx. 1989. Spatial variability of soil chemical properties in grazed pastures. *Soil Science Society of America Journal* 53: 784–789.

(91) Whitehead, D. C. 1995. *Grassland Nitrogen.* CAB International, Oxon, England.

(92) Wilkinson, S. R., J. A. Stuedemann, and D. P. Belesky. 1989. Soil potassium distribution in grazed K-31 tall fescue pastures as affected by fertilization and endophytic fungus infection level. *Agronomy Journal* 81: 508–512.

(93) Wohl, N. E., and R. F. Carline. 1996. Relations among riparian grazing, sediment loads, macro-invertebrates, and fishes in three central Pennsylvania streams. *Canadian Journal of Fisheries and Aquatic Science* 53: 260–266.

CHAPTER 6

(1) Amor, R. L. 1972. *Ecology of Blackberry in Victoria.* Pamphlet No. 35. Verminous and Noxious Weeds Destruction Board. Victoria, Australia. pp. 1–4.

(2) Andow, D. A. 1991. Vegetational diversity and arthropod population response. *Annual Review of Entomology* 36: 561–586.

(3) Batten, G. J. 1979. Controlling scrub weeds with goats. *New Zealand Weed and Pest Control Society* 32: 292–296.

(4) Blossey, B., D. Schroeder, S. D. Hight, and R. A. Malecki. 1994. Host specificity and environmental impact of two leaf beetles *(Galerucella calmariensis* and *G. pusilla)* for biological control of purple loosestrife *(Lythrum salicaria). Weed Science* 42: 134–140.

(5) Blossey, B., D. Schroeder, S. D. Hight, and R. A. Malecki. 1994. Host specificity and environmental impact of the weevil *Hylobius transversovittatus,* a biological control agent of purple loosestrife *(Lythrum salicaria). Weed Science* 42: 128–133.

(6) Blume, R. R. 1985. A checklist, distributional record, and annotated bibliography on the insects associated with bovine droppings on pasture in America north of Mexico. *Southwestern Entomology* (Suppl. #9). 55 pp.

(7) Bornemissza, G. F. 1969. The Australian dung beetle research unit in Pretoria. *South African Journal of Science* 75: 257–260.

(8) Bornemissza, G. F. 1976. The Australian dung beetle project 1965–1975. *Australian Meat Research and Communications Review* 30. 32 pp.

(9) Bosworth, S. C., C. S. Hoveland, and G. A. Buchanan. 1985. Forage quality of selected cool season weed species. *Weed Science* 34: 150–154.

(10) Bosworth, S. C., C. S. Hoveland, G. A. Buchanan, and W. A. Anthony. 1980. Forage quality of selected warm season weed species. *Agronomy Journal* 72: 1050–1054.

(11) Bowes, G. G., and A. G. Thomas. 1978. Longevity of leafy spurge seeds in the soil following various control programs. *Journal Range Management* 31: 137–140.

(12) Bright, D. E. 1994. Revision of the genus *Sitona* (Coleoptera: Curculionidae) of North America. *Annals of the Entomological Society of America* 87: 277–306.

(13) Brussard, L. 1998. Soil fauna, guilds, functional groups and ecosystem processes. *Applied Soil Ecology* 9: 123–135.

(14) Byers, R. A., and G. A. Jung. 1979. Insect population of forage grasses: Effects of nitrogen and insecticides. *Environmental Entomology* 8: 11–18.

(15) Byers, R. A., and W. C. Templeton, Jr. 1988. Effects of sowing date, placement of seed, vegetation suppression, slugs and insects upon establishment of no-till alfalfa in orchardgrass sod. *Grass and Forage Science* 43: 279–289.

(16) Byers, R. A., J. W. Neal, Jr., J. H. Elgin, Jr., K. R. Hill, J. E. McMurtrey III, and J. Feldmesser. 1977. Systemic insecticides with spring-seeded alfalfa for control of potato leafhopper. *Journal of Economic Entomology* 70: 337–340.

(17) Capinera, J. L., and T. S. Sechrist. 1982. Grasshopper (Acrididae)-host plant associations: response of grasshopper populations to cattle grazing intensity. *Canadian Entomology* 114: 1055–1062.

(18) Casler, M. D., and G. A. Pederson. 1996. Host resistance/tolerance and its deployment, pp. 475–507, In: S. Chakraborty, K. T. Leath, R. A. Skipp, G. A. Pederson, R. A. Bray, G. C. M. Latch, and F. W. Nutter, Jr. (ed.). *Pasture and Forage Crop Pathology.* American Society of Agronomy, Crop Science Society of America, Soil Science Society of America, Madison, WI.

(19) Combs, R. L. 1988. New beetle species help clean up pastures. *Research Highlights,* Vol. 51, No. 1. Mississippi Agricultural and Forestry Experimental Station.

(20) Culik, N. P. 1995. *Effects of Pasture Management Factors on Red Clover Root and Stand Characteristics and Observations of Seasonal Crawling Activity of Root Feeding Insect Pests of Red Clover.* Ph.D. dissertation, West Virginia University. 169 pp.

(21) Curry, J. P. 1994. *Grassland Invertebrates: Ecology, Influence on Soil Fertility, and Effects on Plant Growth.* Chapman & Hall, London.

(22) DeClerck-Floate, R. 1997. Cattle as dispersers of hound's tongue on rangeland in southeastern British Columbia. *Journal of Range Management* 50: 239–243.

(23) Drea, J. J. 1966. Studies of *Aleochara tristis* (Coleoptera: Staphylinidae), a natural enemy of the face fly. *Journal of Economic Entomology* 59: 1368–1373.

(24) Drummond, R. O., G. Lambert, H. E. Smalley, Jr., and C. E. Terrill. 1981. Estimated losses of livestock pests, pp.111–127, In: D. Pimentel (ed.). *CRC Handbook of Pest Management in Agriculture,* Vol. 1. CRC Press, Inc., Boca Raton, FL.

(25) Ecale, C. L., and E. L. Backus. 1995. Mechanical and salivary aspects of potato leafhopper probing in alfalfa stems. *Entomologia Experimentalis et Applicata* 77: 121–132.

(26) Ferguson, C. M., B. I. P. Barratt, and P. A. Jones. 1988. Control of the grey field slug *(Deroceras reticulatum* Mueller) by stock management prior to direct-drilled pasture establishment. *Journal of Agricultural Science* (Cambridge) 111: 443–449.

(27) Fincher, G.T. 1981. The potential value of dung beetles in pasture ecosystem. *Journal of the Georgia Entomological Society* 16 (Suppl.): 316–333.

(28) Fincher, G.T. 1986. Importation, colonization, and release of dung burying scarabs. pp. 69–76, In: R. S. Patterson and D. A. Rutz (ed.). *Biological Control of Muscoid Flies.* Miscellaneous Publication of the Entomological Society of America, No. 61.

(29) Fincher, G.T. 1990. Biological control of dung breeding flies: Pests of pastured cattle in the United States, pp. 137–151, In: D. A. Rutz and R. S. Patterson (ed.). *Biological Control of Arthropods Affecting Livestock and Poultry.* Westview, Boulder, CO.

(30) Fincher, G.T. 1991. Sustained-release bolus for horn fly (Diptera: Muscidae) control: Effects of methoprene and diflubenzuron on some non-target species. *Environmental Entomology* 20: 77–82.

(31) Fincher, G.T. 1992. Injectable ivermectin for cattle: Effects on some dung-inhabiting insects. *Environmental Entomology* 21: 871–876.

(32) Fishel, F. 2000. Plants poisonous to livestock. *Agricultural MU Guide,* Missouri Extension, University of Missouri, Columbia.

(33) Gardner, F. P., R. B. Pearce, and R. L. Mitchell. 1985. *Physiology of Crop Plants.* Iowa State University Press, Ames. pp. 290–291.

(34) Gillard, P. 1967. Coprophagous beetles in the pasture ecosystem. *Journal of the Australian Institute of Agricultural Science* 33: 30–34.

(35) Grant, J. F., K.V. Yeargan, B. C. Pass, and J. C. Parr. 1982. Invertebrate organisms associated with alfalfa seedling loss in complete-tillage and no-tillage sowings. *Journal of Economic Entomology* 75: 822–826.

(36) Groppe, K. 1992. Gymnetron antirrhini *Paykull (Coleoptera: Curculionidae), a Candidate for Biological Control of Dalmatian Toadflax in North America.* International Institute of Biological Control, European Station Final Report. 22 pp.

(37) Hardin, J. W. 1973. *Stock-Poisoning Plants of North Carolina.* Bulletin No. 414. Agricultural Experiment Station, North Carolina State University, Raleigh.

(38) Harris, R. L., and R. R. Blume. 1986. Beneficial insects inhabiting bovine droppings in the United States. pp. 10–15, In: R. S. Patterson and D. A. Rutz (ed.). *Biological Control of Muscoid Flies.* Miscellaneous Publication of the Entomological Society of America, No. 61.

(39) Hartley, M.J., L.A. Lyttle, and A. I. Popay. 1984. Control of California thistle by grazing management. *Proceedings of the New Zealand Weed Pest Control Conference* 37: 24–27.

(40) Hill, R. J., and D. Folland. 1986. *Poisonous Plants of Pennsylvania.* Pennsylvania Department of Agriculture, Harrisburg, PA.

(41) Hoebeke, E. R., R. A. Byers, M. A. Alonso-Zarazaga, and J. F. Stimmel. 2000. *Ischnopterapion (Chlorapion) virens* (Herbst) (Coleoptera: Curculionidae: Brentidae: Apioninae), a Palearctic clover pest new to North America: recognition features, distribution, and bionomics. *Proceedings of the Entomological Society of Washington* 102: 151–161.

(42) Hölldobler, B., and E. O. Wilson. 1990. *The Ants.* Belknap Press of Harvard Univ. Press, Cambridge.

(43) Hower, A. A., M. A. Quinn, S. D. Alexander, and K. T. Leath. 1995. Productivity and persistence of alfalfa in response to clover root curculio (Coleoptera: Curculionidae) injury in Pennsylvania. *Journal of Economic Entomology* 88: 1433–1440.

(44) Julien, M. H. 1989. Biological control of weeds worldwide: trends, rates of success and the future. *Biocontrol News and Information* 10(4): 299–306.

(45) Kessler, H., and E. U. Balsbaugh, Jr. 1972. Parasites and predators of face fly in east-central South Dakota. *Journal of Economic Entomology* 65: 1636–1638.

(46) Kok, L. T. 1992. *Biological Control of Musk and Plumeless Thistles.* Virginia Cooperative Extension Publication 444–019. pp. 1–8.

(47) Kronberg, S. L., and J. W. Walker. 1993. Ruminal metabolism of leafy spurge in sheep and goats: a potential explanation for differential foraging of spurge by sheep, goats, and cattle. *Journal of Chemical Ecology* 19: 2007–2017.

(48) Leath, K. T., Erwin, D. C., and Griffin, G. D. 1988. Diseases and nematodes. pp. 621–670, In: A. A. Hanson, D. K. Barnes, and R. R. Hill, Jr. (ed.), *Alfalfa and Alfalfa Improvement.* American Society of Agronomy, Crop Science Society of America, Soil Science Society of America. Madison, WI.

(49) Leath, K. T., Griffin, G. D., Onsager, J. A., and Masters, R. A. 1996. Pests. pp. 193–228, In: L. E. Moser, D. R. Buxton, and M. D. Casler (ed.). *Cool-Season Forage Grasses.* American Society of Agronomy, Crop Science Society of America, Soil Science Society of America, Madison, WI.

(50) Leath, K.T., R. E. Welty, R. G. Pratt, and R. M. Sonoda. 1996. Pasture/forage crops and diseases in the United States. pp. 33–58, In: S. Chakraborty, K. T. Leath, R. A. Skipp, G. A. Pederson, R. A. Bray, G. C. M. Latch, and F. W. Nutter, Jr. (ed.). *Pasture and Forage Crop Pathology.* American Society of Agronomy, Crop Science Society of America, Soil Science Society of America, Madison, WI.

(51) Luginbuhl, J. M., T. E. Harvey, J. T. Green, Jr., M. H. Poore, and J. P. Mueller. 1999. Use of goats as biological agents for the renovation of pastures in the Appalachian region of the United States. *Agroforestry* 44: 241–252.

(52) Magadlela, A. M., M. E. Dabaan, W. B. Bryan, E. C. Prigge, J. G. Skousen, G. E. D'souza, B. L. Arbogast, and G. Flores. 1995. Brush clearing on hill land pastures with sheep and goats. *Journal of Agronomy and Crop Science* 174: 1–8.

(53) May, R. M. 1975. Patterns of species abundance and diversity. pp. 81–120, In: M. L. Cody, and J. M. Diamond (ed.). *Ecology and Evolution of Communities.* Belknap Press, Cambridge, MA.

(54) McClay, A.S. 1992. Effects of *Brachypterolus pulicarius* (L.) (Coleoptera: Nitidulidae) on flowering and seed production of common toadflax. *Canadian Entomology* 124: 631–636.

(55) Olson, B.E. 1999. Grazing and weeds. pp. 85–96, In: R. L. Sheley, and J. K. Petroff (ed.). *Biology and Management of Noxious Rangeland Weeds.* Oregon State University Press, Corvallis.

(56) Olson, B. E., R. T. Wallander, and R. W. Kott. 1997. Recovery of leafy spurge seed from sheep. *Journal of Range Management* 50: 10–15.

(57) Parlevliet, J. E. 1993. What is durable resistance? A general outline. pp. 23–39, In: T. H. Jacobs and J. E. Parlevliet (ed.). *Durability of Disease Resistance.* Kluwer Academic Publishers, Boston, MA.

(58) Pennypacker, B. W., and Hall, M. H. 1998. Is *Sclerotinia sclerotiorum* a concern in spring establishment of alfalfa following soybean? *Phytopathology* 88: S136.

(59) Peters, E. J. 1973. *Control of Pasture Weeds by Increasing Competition from Forages.* Abstr. 59, Weed Science Society of America.

(60) Popay, I., and R. Field. 1996. Grazing animals as weed control agents. *Weed Technology* 10: 217–231.

(61) Powell, R. D., and J. H. Myers. 1988. The effect of *Sphenoptera jugoslavica* Obenb. (Col., Burprestidae) on its host plant *Centaurea diffusa* Lam. (Compositae). *Journal of Applied Entomology* 106: 25–45.

(62) Price, P. W. 1984. *Insect Ecology,* 2nd ed. John Wiley & Sons, New York.

(63) Provenza, F. D. 1995. Postingestive feedback as an elementary determinant of food selection and intake in ruminants. *Journal of Range Management* 48: 2–17.

(64) Richardson, P. Q., and R. J. Richardson. 2000. Dung beetles and their effect on soil. *Ecological Restoration* 18:116.

(65) Schreiber, M. M., and L. R. Oliver. 1969. *Microenvironment of Weed Competition in Alfalfa Establishment.* Abstr. 156, Weed Science Society of America.

(66) Schreiber, M. M. 1967. Effect of density and control of Canada thistle on production and utilization of alfalfa pasture. *Weeds* 15: 138–142.

(67) Smith, C. M. 1989. *Plant Resistance to Insects, a Fundamental Approach.* John Wiley & Sons, New York.

(68) Smith, L. M., F. W. Ravlin, L. T. Kok, and W. T. Mays. 1984. Seasonal model of the interaction between *Rhinocyllus conicus* (Coleoptera: Curculionidae) and its weed host, *Cardus thoermeri* (Campanulatae: Asteraceae). *Environmental Entomology* 13: 1417–1426.

(69) Sobhian, R., G. Campobasso, and P. H. Dunn. 1992. A contribution to the biology of *Bangasternus fausti* (Col., Curculionidae), a potential biological control agent of diffuse knapweed, *Centaurea diffusa,* and its effect on the host plant. *Entomophaga* 37: 171–179.

(70) Story, J. M., Boggs, K.W., and Nowierski, R. M. 1989. The effect of two introduced seedhead flies on spotted knapweed. *Montana Agriculture Research,* Winter: 14–17. Montana State University, Bozeman, MT.

(71) Stuteville, D. L., and D. C. Erwin (ed.). 1990. *Compendium of Alfalfa Diseases.* APS Press, St. Paul, Minnesota.

(72) Tashiro, H. 1987. *Turfgrass Insects of the United States and Canada.* Comstock Publishing Assoc., Cornell University Press, Ithaca.

(73) Underwood, J. F., M. M. Loux, J. W. Amrine, and W. B. Bryan. 1996. *Multiflora Rose Control.* Ohio State University Bulletin 857. Columbus, OH.

(74) Vengris, J., M. Drake, W. G. Colby, and J. Bart. 1953. Chemical composition of weeds and accompanying crop plants. *Agronomy Journal* 45: 213–218.

(75) Wall, R., and L. Strong. 1987. Environmental consequences of treating cattle with the anti-parasitic drug ivermectin. *Nature* 327: 418–421.

(76) Waterhouse, D. F. 1974. The biological control of dung. *Scientific American* 230: 101–109.

(77) Welton, F. A., V. H. Morris, and A. J. Hartzler. 1929. *Organic Food Reserves in Relation to the Eradication of Canada Thistle.* Ohio Agricultural Experiment Station Bulletin 441.

(78) White, I. M., and K. Marquardt. 1989. A revision of the genus *Chaetorellia hendel* (Diptera: Tephritidae) including a new species associated with spotted knapweed, *Centaurea maculosa* Lam. (Asteraceae). *Bulletin of Entomological Research* 79: 453–487.

(79) Wilson, F., and C. B. Huffaker. 1976. The philosophy, scope and importance of biological control. pp. 3–15, In: C. B. Huffaker and P. S. Messenger (ed.). *Theory and Practice of Biological Control.* Academic Press, New York.

(80) Wood, G.M. 1987. Animals for biological brush control. *Agronomy Journal* 79: 319–321.

CHAPTER 7

(1) Badger, T. H. 1983. *Evaluation of Sod-Seeded Pasture Renovation.* M.S. thesis. University of Maryland, College Park.

(2) Brown, B. A., A. M. Decker, M. A. Sprague, H. A. MacDonald, M. R. Teel, and J. B. Washko. 1960. *Band and Broadcast Seeding of Alfalfa-Bromegrass in the Northeast.* Maryland Agricultural Experiment Station Bulletin A-108 and Northeast Regional Publication 41.

(3) Byers, R. A., and W. C. Templeton, Jr. 1988. Effects of sowing date, placement of seed, vegetation suppression, slugs and insects upon establishment of alfalfa in orchardgrass sod. *Grass and Forage Science* 43: 279–289.

(4) Campbell, M. H. 1976. Effect of timing of glyphosate and 2,2-DPA application on establishment of surface-sown pasture species. *Australian Journal of Experimental Agriculture and Animal Husbandry* 16: 491–499.

(5) Decker, A. M., and R. F. Dudley. 1976. Minimum tillage establishment of five forage species using five sod-seeding units and two herbicides. pp. 140–146, In: J. Luchok, J. D. Cawthon, and M. J. Breslin (ed.). *Hill Lands. Proceedings of an International Symposium.* West Virginia University Books, Morgantown.

(6) Decker, A. M., H. J. Retzer, F. G. Swain, and R. F. Dudley. 1969. *Midland Bermudagrass Forage Production Supplemented by Sod-Seeded Cool-Season Annual Forages.* Maryland Agricultural Experiment Station Bulletin 484.

(7) Decker, A. M., J. H. Vandersall, and N. A. Clark. 1976. Pasture renovation with alternate row sod-seeding of different legume species. pp. 146–149, In: J. Luchok, J. D. Cawthon, and M. J. Breslin (ed.). *Hill Lands. Proceedings of an International Symposium.* West Virginia University Books, Morgantown.

(8) Fink, R. J., and O. H. Fletchall. 1963. Forage crop establishment in soil containing atrazine or simazine residues. *Weeds* 11: 81–83.

(9) Hannaway, D. B., and P. E. Shuler. 1993. Nitrogen fertilization in alfalfa production. *Journal of Production Agriculture* 6: 80–85.

(10) Koch, D. W., and G. O. Estes. 1986. Liming rate and method in relation to forage establishment—Crop and soil chemical responses. *Agronomy Journal* 78: 567–571.

(11) Lowder, S. W., and J. B. Weber. 1982. Atrazine efficacy and longevity as affected by tillage, liming, and fertilizer type. *Weed Science* 30: 273–280.

(12) Mueller-Warrant, G. W., and D. W. Koch. 1980. Establishment of alfalfa by conventional and minimum-tillage seeding techniques in a quackgrass dominant sward. *Agronomy Journal* 72: 884–889.

(13) Sheard, R. W. 1980. Nitrogen in the P band for forage establishment. *Agronomy Journal* 72: 89–97.

(14) Sheard, R. W., G. J. Bradshaw, and D. L. Massey. 1971. Phosphorus placement for the establishment of alfalfa and bromegrass. *Agronomy Journal* 63: 922–927.

(15) Smith, E. M., T. H. Taylor, J. H. Casada, and W. C. Templeton. 1973. Experimental grassland renovator. *Agronomy Journal* 65: 506–508.

(16) Sprague, M. A., M. M. Hoover, Jr., M. J. Wright, H. A. MacDonald, B. A. Brown, A. M. Decker, J. B. Washko, V. G. Sprague, and K. E. Varney. 1963. *Seedling Management of Grass-Legume Associations in the Northeast.* New Jersey Agricultural Experiment Station Bulletin 804 and Northeast Regional Publication 42.

(17) Taylor, T. H., and L. T. Jones, Jr. 1982. Persistence and productivity of sod-seeded legumes compared with nitrogen fertilized grass sod. p. 129, In: *Agronomy Abstracts.* American Society of Agronomy, Madison, WI.

(18) Taylor, T. H., E. M. Smith, and W. C. Templeton. 1969. Use of minimum tillage and herbicide for establishing legumes in Kentucky bluegrass *(Poa pratensis* L.) swards. *Agronomy Journal* 61: 761–766.

(19) Undersander, D. 1999. *Seeding Rate of Different Alfalfa Seed Lots.* University of Wisconsin Cooperative Extension. Available at http://www.uwex.edu/ces/forage/pubs/seedrate.html.

(20) Vough, L. R., A. M. Decker, and R. F. Dudley. 1982. Influence of pesticide, fertilizers, row spacings, and seeding rates on no-tillage establishment of alfalfa. pp. 547–550, In: J. A. Smith and V. W. Hays (ed.). *Proceedings of the 14th International Grassland Congress.* Westview Press, Boulder, CO.

(21) Ward, C.Y., and R. E. Blaser. 1961. Effect of nitrogen fertilizer on emergence and seedling growth of forage plants and subsequent production. *Agronomy Journal* 53: 115–120.

(22) Welty, L. E., R. L. Anderson, R. H. Delaney, and P. F. Hansleigh. 1981. Glyphosate timing effects on establishment of sod-seeded legumes and grasses. *Agronomy Journal* 73: 813–817.

Other Books from NRAES

The books below can be ordered from NRAES (Natural Resource, Agriculture, and Engineering Service). Complete book descriptions are posted on the NRAES web site. Before ordering, contact NRAES for current prices and shipping and handling charges, or for a free catalog.

NRAES, Cooperative Extension
PO Box 4557
Ithaca, New York 14852-4557
Phone: (607) 255-7654
Fax: (607) 254-8770
E-mail: NRAES@CORNELL.EDU
Web site: WWW.NRAES.ORG

Features

- 32 papers
- 70+ tables
- 50+ illustrations

Dairy Calves and Heifers: Integrating Biology and Management

NRAES-175 • 398 pages • 2005
ISBN 0-935817-97-2

This conference proceedings provides information on managing nutrition and health to reduce sickness and age at first calving. Includes cost analysis of replacement programs and housing considerations. Intended for dairy producers, their advisors, educators, and agribusiness.

Features

- Decision trees
- 38 illustrations
- 26,000+ sold

First on the Scene

NRAES-12 • 46 pages • 1989
ISBN 0-935817-06-9

Teaches good decision-making procedures to apply between accident discovery and rescue personnel arrival. Includes decision-and-action diagrams for accidents involving machinery, grain and silage storage facilities, chemicals, and electrocution. Intended for educators, farm families, and farm workers.

Features

- 24 color photos
- 24 illustrations
- 17 tables

Field Guide to On-Farm Composting

NRAES-114 • 118 pages • 1999
ISBN 0-935817-39-5

Discusses site considerations, raw materials, equipment, recipe making, troubleshooting, mortality composting, and compost utilization. Includes equipment capacity table and material properties. Intended for compost system managers and educators.

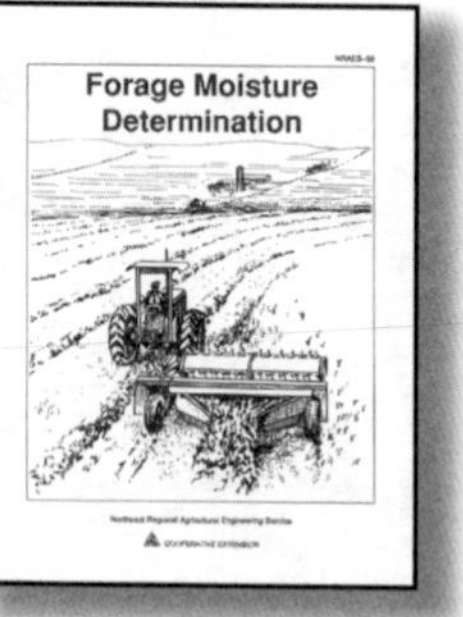

Features

- 11 illustrations
- Equations
- Summary table

Forage Moisture Determination

NRAES-59 • 28 pages • 1993

The moisture content of stored forage is crucial to successful preservation. This book describes on-farm and laboratory methods of determining moisture content as well as techniques for taking representative samples for testing.

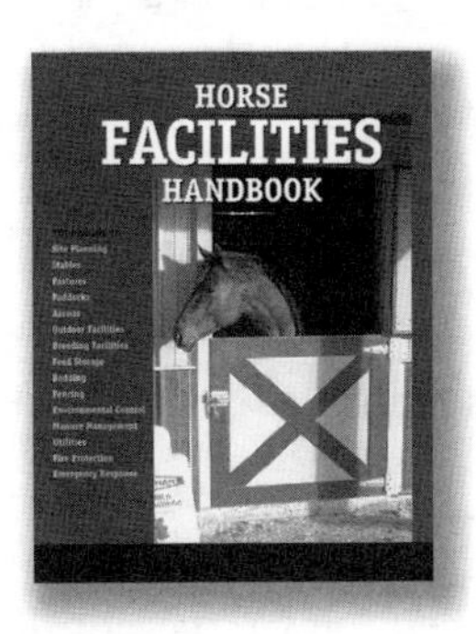

Horse Facilities Handbook

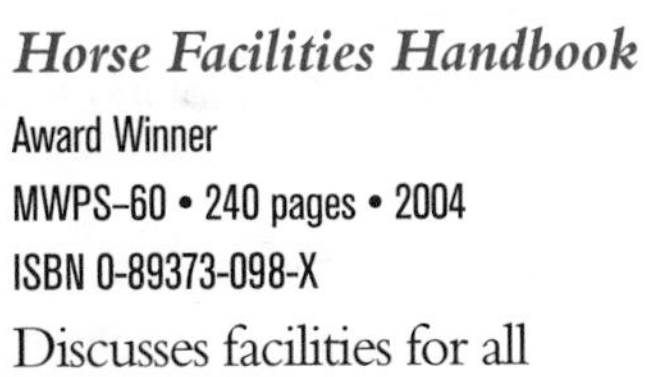

Award Winner

MWPS-60 • 240 pages • 2004
ISBN 0-89373-098-X

Discusses facilities for all phases of breeding, care, and control. Topics include site planning; construction of stables, paddocks, and other indoor and outdoor facilities; environmental control; manure management; bulk feed and bedding storage; fencing; utilities; fire protection; and emergency response planning.

Features

- 111 drawings
- 50+ photos
- Index

Silage and Hay Preservation

Award Winner

NRAES-5 • 53 pages • 1990
ISBN 0-935817-47-6

This book helps farmers conserve the digestible fiber, protein, and energy of forage and maintain it in a form that animals can use efficiently. Topics covered are the biology of silage preservation, preservation of hay, and additives for silage and hay preservation.

Features

- 15 charts/graphs
- 47 tables
- Equations

Managing and Marketing for Pasture-Based Livestock Production

NRAES-174 • 114 pages • 2006
ISBN-13: 978-0-935817-99-7
ISBN-10: 0-935817-99-9

Focuses on the framework of management decisions around which a forage-livestock production system is developed. Managers will learn to develop a vision statement; formulate goals and plans; inventory and allocate farm resources; and move commercial feeder cattle, dairy, and hay products into the marketplace.

Features

- 35+ figures/tables
- Sample budgets
- Commodity standards

Silage for Dairy Farms: Growing, Harvesting, Storing, and Feeding

NRAES-181 • 450 pages • 2006
ISBN-13: 978-1-933395-06-7
ISBN-10: 1-933395-06-0

Focuses on silage management for improved profitability and reduced environmental impact. Topics include managing nutrients through forages, producing crops for silage, sources and management of mycotoxins, harvesting forages, silage storage strategies, and silage feeding considerations. Intended for dairy producers, their advisors, educators, and agribusiness. Proceedings of a 2006 conference.

Features

- 37 papers
- 86 illustrations
- 126 tables

About NRAES

NRAES, the Natural Resource, Agriculture, and Engineering Service, is a not-for-profit program dedicated to assisting land grant university faculty and others in increasing the public availability of research- and experience-based knowledge. NRAES is sponsored by fourteen land grant universities in the eastern United States (see map below). We receive administrative support from Cornell University, the host university.

When you buy books from NRAES, you are helping to improve the accessibility of land grant university knowledge. While 15% of NRAES' annual income is provided by member universities, the funds to publish new books and coordinate new conferences come from our customers through book sales, conference registrations, and occasional project-specific grants.

NRAES publishes practical books of interest to fruit and vegetable growers, landscapers, dairy and livestock producers, natural resource managers, soil and water conservation district staff, consumers, landowners, and professionals interested in agricultural waste management and composting. NRAES books are used in cooperative extension programs, in college courses, as management guides, and for self-directed learning.

NRAES publishes two types of books: peer-reviewed books and conference proceedings. Our peer-reviewed books are evaluated prior to publication for technical accuracy and usefulness to the intended audience. The reviewers may include university faculty, extension educators, potential users, and interested persons from government and agribusiness. Conference proceedings are not peer-reviewed. However, the authors of papers presented at NRAES-sponsored conferences are chosen for their recognized expertise. NRAES also distributes some videos related to waste management and fruit and vegetable production.

NRAES was started in 1974 and was originally known as the Northeast Regional Agricultural Engineering Service. In 1987, with encouragement from member university extension directors, NRAES began offering its services to faculty from all disciplines at member universities. In 1998, Virginia Polytechnic Institute and State University joined the original thirteen member universities. Our name was changed in 1998 to reflect the expansion of NRAES beyond the Northeast and the broadening scope of our books and conferences. Contact us for more information or a free book catalog.

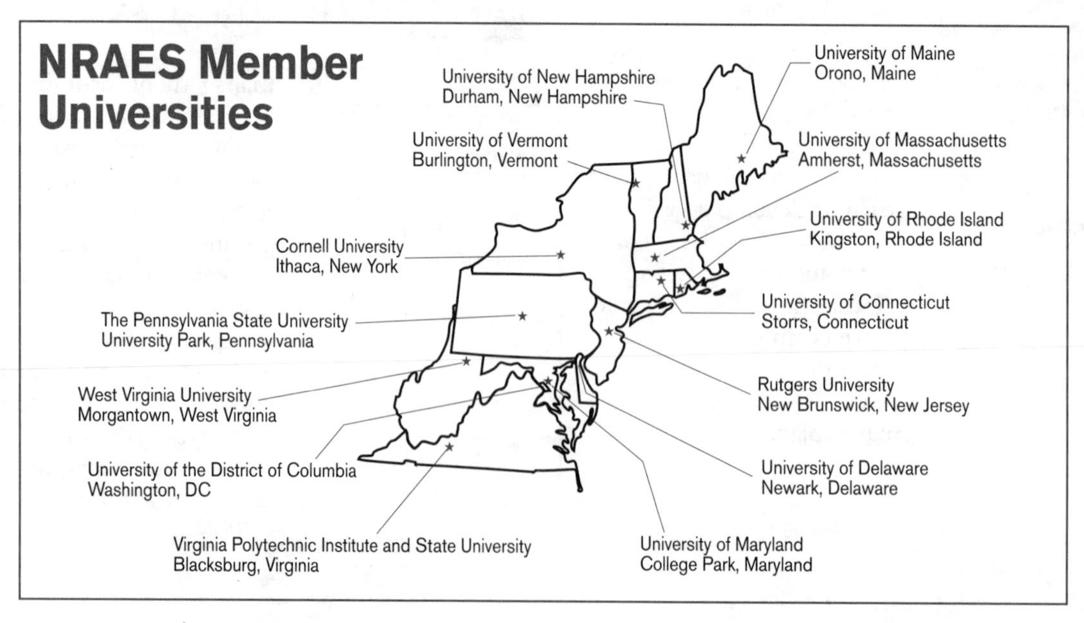

NRAES
Cooperative Extension
PO Box 4557
Ithaca, New York 14852-4557

Phone: (607) 255-7654
Fax: (607) 254-8770
E-mail: NRAES@CORNELL.EDU
Web site: WWW.NRAES.ORG

Marty Sailus, NRAES Director
Jeffrey S. Popow, NRAES Managing Editor